LIFE SCIENCES AND SPACE RESEARCH XIII

COSPAR

LIFE SCIENCES AND SPACE RESEARCH XIII

Proceedings of the Open Meeting of the Working Group on Space Biology of the Seventeenth Plenary Meeting of COSPAR

São Paulo, S. P., Brazil — June 1974

Organized by

THE COMMITTEE ON SPACE RESEARCH — COSPAR

and

THE "INSTITUTO DE PESQUISAS ESPACIAIS — INPE" OF BRAZIL

Edited by

P. H. A. SNEATH

AKADEMIE-VERLAG · BERLIN

1975

Executive Editor:
Dr. A. C. Stickland

Library of Congress Catalog Card Number 63—6132

202 · 100/474/75
Gesamtherstellung: VEB Druckhaus „Maxim Gorki", 74 Altenburg
Bestellnummer: 762 128 5 (3060/XIII) · LSV: 1305, 1495
Printed in GDR

Preface

The papers in this volume, the thirteenth in the series of papers on space biology were presented to the 17th Plenary Meeting of the Committee on Space Research (COSPAR). It contains the proceedings of the Life Sciences sessions of COSPAR held in São Paulo in June 1974. These sessions were dedicated to the memory of Professor Wolf Vishniac. An obituary was included in the last volume (XII) of the series.

The effects of weightlessness and of greatly increased acceleration continue to excite biological interest. These aspects of space biology are well represented by contributions in the present volume. There is growing evidence that very significant biological changes may be involved. The other main area reporting results of flight experiments is radiation biology, and we are now beginning to benefit from the meticulous engineering in space vehicles that allows sophisticated analysis of the action of cosmic rays, including some evidence on the site of action in the living cell.

A section of the contributions deals with the biology of extreme conditions, and documents a number of factors that affect growth and survival in very harsh environments. Further interest in planetary quarantine is also evinced by several contributions to this field, in which the Jovian planets and their satellites are now assuming significance.

The editor would like to express his grateful thanks to Dr. A. C. Stickland for her work in technical editing and in preparing this volume for press. She has overcome many problems associated with the large correspondence involved and the need to keep to a tight printing schedule.

P. H. A. Sneath
27 September 1974

Contents List

Gravitational Biology: Continuity and Proportionality of Acceleration Effects

Life Sciences and Space Research XIII — Akademie-Verlag, Berlin 1975

HUMAN SENSITIVITY TO GRAVITY (ON THE PROBLEM OF GRAVIPREFERENDUM)

E. B. SHULZHENKO

Moscow, USSR

Experiments were carried out in a centrifuge with an arm of 7.25 m, the cabin being equipped with a self-contained system of control. During an exposure to head-to-feet accelerations ($+g_z$) of 2 g and an onset rate of 0.1 $g\,s^{-1}$ the test subject controlled the centrifugation himself and selected the acceleration values he was instructed to attain. The subjects were shown to be capable of assessing integrally the value of accelerations and estimating it with an error of 0.1—0.12 g.

During an exposure to chest-to-back accelerations ($+g_x$) limited to a value of 12 g and a gradient of 0.3 $g\,s^{-1}$ the ability to select actively the rate of an acceleration increase was studied. Repeated centrifugations resulted in the formation and strengthening of the skill of maintaining a physiologically optimal rate of an acceleration increase. Onset rates were determined for every level of acceleration. When calculated as a mean, physiologically optimal values of the onset rate for 12 g were 0.16—0.18 $g\,s^{-1}$.

The findings give insight into important properties of the physiological structure of "gravipreferendum", which is the capability of man to differentiate between gravity levels and to select actively optimal rates of an increase in acceleration.

Almost two hundred years had passed after Isaac Newton discovered the law of gravity (1687) before a hypothesis about the role of gravity in the life of terrestrial organisms was advanced. This hypothesis was put forth by Konstantin E. Tsiolkovsky who discussed problems of the relationship between the size of an organism and its environment in his paper "Biology of Dwarfs and Giants" (1882). Gravity influences the development, structure and function of different systems of an organism. The correlation between the final size of a body and the gravity field has been demonstrated experimentally. Increased gravity inhibits size and gain in weight of different animals, affects substantially the development of the musculoskeletal system, smooth musculature of viscera, etc. [1—5]. An analysis of the literature data shows convincingly the predominant effect of gravity on the development of specific formations for which deformation is an adequate stimulation. According to modern concepts about interoreceptors [6], these formations are functionally united by the term "mechanoreceptors" whose excitation is related to the deforming influences. For example, such formations in the circulation system include many "pressoreceptors" of the vascular bed whose distension (induced by mechanical deformation) gives rise to the afferent pathway in the neuro-reflex regulation of blood pressure. This is particularly true of the reflexogenic zone in the cardiovascular system (carotid sinus, aorta, etc.), which are responsible for the maintenance of haemodynamic parameters at an adequate level during changes of the vector of the gravity field.

The effect of gravity is related to the tone of anti-gravity muscles and equilibrium centres having a complex system of statokinetic reflexes. Central and peripheral formations responsible for the orientation of the body within the gravity field include the cerebellar system, vestibular structures, optic perception and the proprioceptive function of the motor analyser [7—10].

In space flights of long duration changes in the level of gravity become in certain situations a gravity barrier. One of the most important problems of space biology and medicine is to know how to overcome this barrier. This problem can be defined as the subject of gravity physiology (O. G. Gazenko).

One of the most important problems of the theory and practice of this branch of science is to estimate man's capacity to measure changes in gravity from a

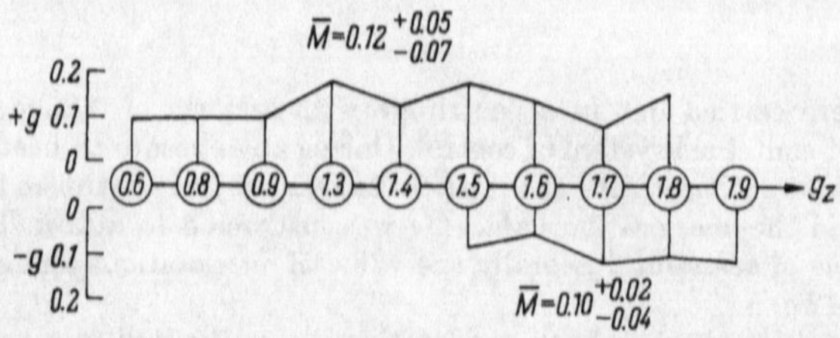

Fig. 1. Value of an error in the assessment of human gravity sensitivity and its dependence on the acceleration value.

constant value. There is a report in the literature that indicates that organisms are sensitive to the gravity level [11]; when exposed to parabolic centrifugation at a steady angular velocity, animals chose the minimal acceleration moving toward the minimal radius. The animals regarded accelerations of higher than 1 g as unpleasant. Using special methods of controlling angular velocity, the authors carried out experiments on monkeys and demonstrated that they avoided an increase in acceleration from 1.1 to 2.1 g.

The purpose of our investigations was to measure the level of gravity sensitivity in man, its threshold value and acceleration-dependence, as well as to determine the effect of repeated centrifugation on the active formation and strengthening of the skill to maintain an optimal rate of acceleration increase.

We carried out 90 experiments in which 10 test subjects took part. Increased gravity was simulated by centrifugation in a centrifuge with a radius of 7.25 m. The test subjects used a self-contained system of control from the cabin. They were exposed to head-to-feet ($+g_z$) accelerations. The maximum limits of centrifugation were: value of acceleration 2 g and onset rate of acceleration 0.1 $g\ s^{-1}$. At the beginning of centrifugation the subjects were exposed to acceleration of 0.5—2 g for 3—5 min. All the test subjects held the opinion that this time period was sufficient to perceive a given gravity value. Then the test subjects attempted to reproduce the given value, using the system of manual controls.

The experimental findings demonstrate that man has a high enough capacity to evaluate the gravity level with an error of 0.1 to 0.12 g (Fig. 1). They also show qualitative peculiarities of the gravity sensitivity of man. This is indicated by the fact that up to accelerations of $+1.4\ g$ the error has a plus sign and up to accelerations of 0.9 g it does not exceed 0.1 g. Over 0.9 g the level of the error increases slightly; beginning with an acceleration of 1.5 g the error is within the same range of about one tenth part of g, but its value may have a minus sign. Therefore, the test subjects show a sufficiently high gravity sensitivity, although they evaluate the acceleration value in a different way. With an increase of the gravity level they slightly underestimate the absolute value of accelerations. Over an acceleration of 1.8 g the error has a minus sign. Thus, the human body is capable of sensing the level of increased gravity up to 2 g with a high degree of accuracy.

From the physiological point of view this phenomenon can be interpreted as an adaptation of an organism in the broad sense of adaptive reactions useful for a given environment. The following mechanisms responsible for the equilibration of the body and the environment should be noted:

(i) Physiological adaptation, as the capacity to reduce excitability under the influence of a stimulus.

(ii) Functional mobility of receptors, as the capacity to vary the number of the acting receptors in response to qualitative and quantitative fluctuations of the stimulus.

(iii) Function of the body protection, as numerous adaptive changes in the functional system protecting a constant level of its activity.

Any deviation of a certain parameter from the constant level triggers an immediate mobilization of mechanisms contributing to the return to its constant level. This is based on the process of self-control of physiological functions as a branched interaction of central and peripheral formations that constitute a system with certain physiological properties. These self-controlled systems can be illustrated by systems responsible for arterial pressure, body temperature, oxygen consumption and carbon dioxide production, etc. The mechanisms assure with a good reliability the processes of adaptation to numerous and various influences of the environment. The level of earth gravity makes a certain contribution to the structure of developed, and during evolution strengthened, adaptive reactions of the body and mechanisms for their maintenance. From this point of view gravity sensitivity of man can be associated with the fact that different values of mechanical deformation can be well differentiated by various levels of afferent impulsation from mechanoreceptors, provided an adequate level of excitation in regulatory centres is maintained.

The findings about the human capacity to sense gravity levels have formed the basis of further investigations to study the possibility of the active formation and strengthening of the skill to maintain the physiologically optimal rate of increase of accelerations during centrifugations. From the physical point of view the rate of an increase of acceleration or its onset rate ($\Delta g/\Delta t$) is a measure of a change of its value per time unit and is designated $g\ s^{-1}$. If accelerations are considered as a state of an increased weight, then the onset rate determines the value of body weight increment (ΔP) per time unit. The rate of increment of gravity brings about a different rate of deformation of organs and tissues of the organism, and the development of elastic tensions in the body.

There are data available describing changes in the process of adaptation of the body to accelerations depending upon different rates of increase. Increasing onset rates have been studied during an exposure to head-to-feet accelerations [12—14]. The tolerance to head-to-feet accelerations prior to the development of "blackout" during an exposure to 1.9 g was higher with a slow onset rate (0.07 $g\ s^{-1}$)

than with a rapid onset rate (1 $g\,s^{-1}$) [15]: according to Babushkin [16], an onset rate of 0.2 $g\,s^{-1}$ is optimal. However, publications discussing the selection of an optimal increase of transverse accelerations are scanty. Gauer and Zuidema [17] studied onset rates from 0.05 to 3 $g\,s^{-1}$ during an exposure to chest-to-back accelerations. The authors concluded that the tolerance to accelerations was better with a lower onset rate.

The selection of the onset rates providing optimal tolerance to the chest-to-back accelerations was performed, using the method of active choice of the rate of acceleration increase. By an optimal regime of acceleration increase we mean the period of time required to trigger different levels of control responsible for the formation of adequate physiological reactions of the organism during the transition from one value of acceleration to another. By means of the manual system of control the test subject selected the rate of acceleration. The range of the exposure was limited at the control desk by $+12\,g_x$ and an onset rate of 0.3 $g\,s^{-1}$. Test subjects were exposed to chest-to-back accelerations, the back of the chair being inclined to the vector at an angle of 78°. Test subjects were instructed to carry out centrifugation with a constant increase of the rate of acceleration up to $+12\,g$. The time at the plateau of $+12\,g$ was a constant value of 10 sec. Centrifugation was braked automatically at a rate of 0.2 $g\,s^{-1}$. Four test subjects took part in the experiments; each of them carried out 11 centrifugations with an interval of 6 days between two consecutive experiments. From the 1st to the 9th experiment the test subjects were allowed to follow the pattern of accelerations, using the point indicator without knowing the value of accelerations. During the 10th and 11th centrifugations the indicator was removed in order to measure the skill for the maintenance of an optimal rate of acceleration increase.

An analysis of the experimental data demonstrated that the difference of average values of onset rates of accelerations was statistically not significant during the 10th and 11th centrifugations (without the indicator) compared with the 6th—9th experiments (Table 1).

Table 1

Mean Values of Onset Rates for 4 subjects on 11 successive trials of an Acceleration Increase ($\Delta g/\Delta t$) up to 12 g

Centrifugations	1	2	3	4	5	6	7	8	9	10	11
gs^{-1}	0.13	0.09	0.15	0.14	0.18	0.16	0.17	0.17	0.16	0.17	0.18

This fact may provide evidence that during repeated centrifugations there is an increase in the skill for the maintenance of an optimal rate of an acceleration. This is confirmed by the averaged absolute values of onset rates for every value of acceleration (from 1 to 12 g) during repeated centrifugations (Fig. 2). It should be noted that as centrifugations were repeated the onset rates, beginning with the 6th experiment, became stabilized for each acceleration level. With respect to accelerations of 6 g their absolute values increased as centrifugations were repeated. The development of the controlling skill also finds support from the smaller variability of mean onset rates calculated for each level of acceleration from the 6th to the 10th centrifugation.

An analysis of physiological parameters with respect to the heart rate and respiration rate shows that changes in the organism did not vary qualitatively and were characteristic of chest-to-back accelerations. However, a comparison of the averaged indices during the period of acceleration and plateau in the course of repeated centrifugations revealed certain differences. The differences were typical of all test subjects. Statistical analysis of heart rate variations showed

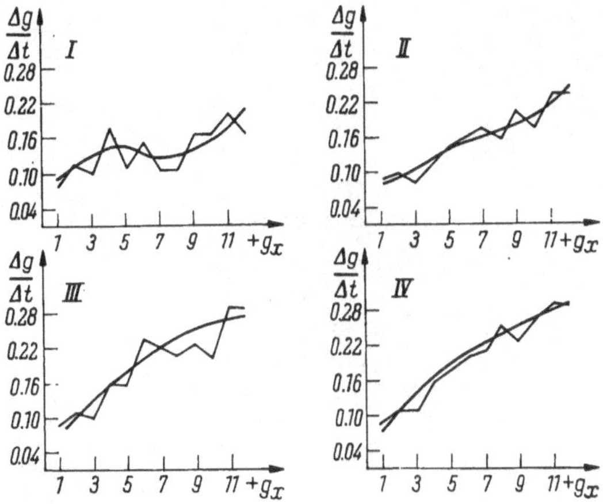

Fig. 2. Dependence of absolute values of average onset rate on repeated centrifugations upon an attainment of $+12\,g$: I, 1st centrifugation; II, 2nd to 5th centrifugation; III, 6th to 9th centrifugation; IV, 10th to 11th centrifugation.

that during the acceleration period as compared with the 1st—3rd centrifugations the heart rate decreased by 6.4% ($P < 0.25$) during the 7th—9th centrifugations and by 12.7% ($P < 0.05$) during the 10th—11th centrifugations (Fig. 3). The heart rate during the plateau also tended to decline ($P < 0.25$).

The respiration rate during the acceleration and plateau periods decreased from centrifugation to centrifugation ($P < 0.25$). The respiratory volume and minute respiratory volume increased, beginning from the 7th centrifugation, and during the 10th—11th centrifugations differed significantly as compared with the changes

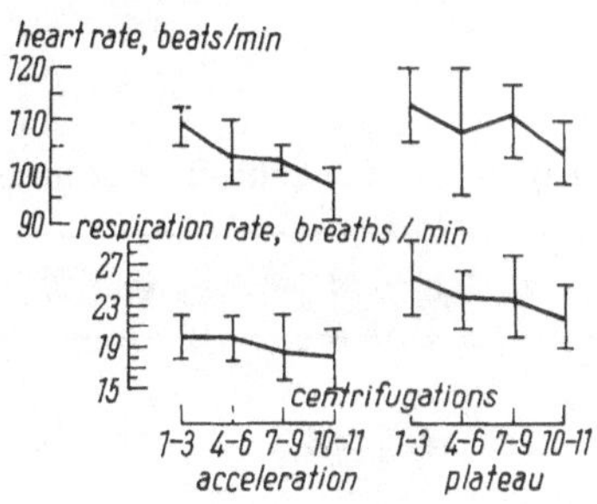

Fig. 3. Changes in the heart rate and respiration rate during repeated centrifugations.

during the 1st—3rd centrifugations ($P < 0.05$) (Fig. 4). This provides evidence that repeated centrifugations, beginning with the 7th exposure, give rise to adaptive processes. With subsequent centrifugations the adaptation is strengthened by an optimal regime of achieving accelerations; this is confirmed by a more rationalized relation of functional changes. This is also supported by the force

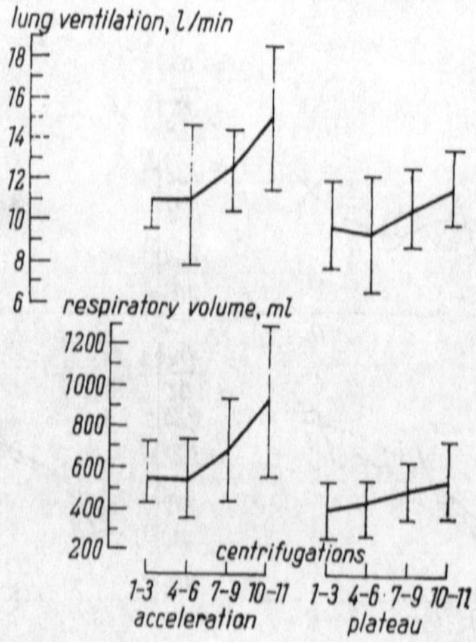

Fig. 4. Lung ventilation and respiration volume during repeated centrifugations.

which the test subject applies to the control stick (Table 2). The difference in force during the 10th—11th centrifugations is lower than the average difference during the 1st—9th centrifugations. However, this difference of force occurs up to accelerations of $+6\,g$ ($P < 0.05$); with higher accelerations this parameter remains unchanged. This may be associated with the fact that at accelerations higher than $+6\,g$ a well pronounced increase in the tonic tension of skeletal muscles (due to the high gravity) occurs. Thus, repeated centrifugations help to develop and strengthen the skill of control; the process develops as an active selection and optimization of acceleration increases up to $+12\,g$. The principle allows estimation of the value of onset rates of increase for each acceleration value, as well as their mean value ($0.16-0.18\,g\,s^{-1}$).

During active selection of acceleration onset rates the rate of their increase

Table 2

Mean Values of the Force applied to the Control Stick for every Acceleration Level

Acceleration g		1	2	3	4	5	6	7	8	9	10	11	12
Force, units	1st to 9th centrifugation	25	29	44	56	67	78	87	106	117	133	143	151
	10th to 11th centrifugation	16	28	40	53	62	76	88	106	117	134	144	152

does not follow a linear relationship; it is characterized by the fact that the greater the acceleration, the higher is the onset rate.

It should be noted that self-controlled systems are characterized by the fact that different impulses in receptors of functional systems are mobilized before an increase in the strength of the deviating factor. Therefore an increasing deviation of the parameter (or a new deviation) brings about a somewhat higher mobilization of the feedback system in the central nervous system. This process of progressive, anticipating mobilization of protective and adaptive mechanisms can be blocked during accelerations that grow extremely rapidly. On the contrary, an extremely low rate of their increase may result in an exhaustion of the adaptive processes.

Our findings illustrate important features in the physiological structure of "gravipreferendum", which is the capacity of man to differentiate accurately the level of gravity, to select actively and to maintain an optimal rate of acceleration increase during repeated centrifugation. This process seems to be based upon the ability to develop and strengthen a new dynamic stereotype in the coordination of intersystemic interactions which can be referred to the category of physiological adaptation. An important part in this process is played by afferent impulsation with a feedback system which assures an optimal state of the regulation of physiological functions in response to stimulation. The gravipreferendum "seems to expand the boundaries within which the organism exists and to weaken its dependence on the environment" [18].

References

[1] V. I. Baranov et al., in: Aviation and Space Medicine, Proc. Conf. Moscow 1963 (p. 56).
[2] A. H. Smith et al., Growth **23**, 97 (1959).
[3] S. R. Briney et al., Proc. Iowa Acad. Sci. **67**, 495 (1960).
[4] F. L. Steel et al., Nature **193**, 583 (1962).
[5] R. R. Burton et al., J. Aerospace Med. **34**, 249 (1963).
[6] V. N. Chernigovsky, Interoreceptors, Medgiz, Moscow 1960.
[7] V. S. Gurfinkel et al., Regulation of Man's Posture, Nauka, Moscow 1965.
[8] E. K. Sepp, History of the Development of the Nervous System of Vertebrates, Medgiz, Moscow 1959.
[9] P. O. Nortington, J. Aviation Med. **5**, 103 (1934).
[10] G. R. Armstrong, J. Brit. Interplan. Soc. **12**, 172 (1953).
[11] F. G. Clark, Amer. J. Astronaut, Aeronaut. **10**, 855 (1971).
[12] E. W. Salzman et al., Circ. Research **4**, 5, 40 (1956).
[13] D. E. Torphy et al., Aerosp. Med. **37**, 52 (1966).
[14] G. H. Cohen et al., Aerosp. Med. **27**, 858 (1969).
[15] R. W. Edelberg et al., J. Aviation Med. **27**, 168 (1956).
[16] V. I. Babushkin, Voyenno-Medizinsky Zh. **8**, 50 (1959).
[17] O. H. Gauer et al., in: Gravitational Stress in Aerospace Medicine, London 1961 (p. 46).
[18] V. V. Parin et al., Essays of Space Physiology, Medgiz, Moscow 1969.

Life Sciences and Space Research XIII — Akademie-Verlag, Berlin 1975

RESPONSE AND ADAPTATION OF BEAGLE DOGS TO HYPERGRAVITY

J. OYAMA

Biomedical Research Division, Ames Research Center, NASA, Moffett Field, Calif., USA

Eight male Beagle dogs, five months old, were centrifuged for three months at progressively increasing loads ranging from 1.5 *g* to 2.5 *g* on a 7.9 m radius centrifuge. Dogs were maintained on the centrifuge, which was run almost continuously, by employing a specially designed automated waste disposal and life support system. Heart rate and deep body temperature were monitored continuously by implant biotelemetry. Initially, centrifuged dogs showed transient decreases in heart rate and body temperature along with changes in their diurnal rhythm patterns. Compared with normal gravity controls, exposed dogs showed a slower growth rate and a reduced amount of body fat. Blood protein, total lipids, cholesterol, calcium, packed cell volume, red blood cell count, and hemoglobin were also decreased significantly. Absolute weights of the leg bones of centrifuged dogs were significantly greater than controls. Photon absorptiometry revealed significant density increases in selective regions of the femur and humerus of centrifuged dogs. In spite of the various changes noted, results from this and other studies affirm the view that dogs can tolerate and adapt to sustained loads as high as 2.5 *g* without serious impairment of their body structure and function.

1. Introduction

Although there have been numerous studies dealing with the effects of long-term hypergravity exposure on small sized animals ranging from mice to chickens [1, 2], there has not been any reported so far on dogs. The present paper describes some of the major changes which have been found in young Beagle dogs which were chronically centrifuged for 12 weeks. This study is part of a gravitational biology program at Ames Research Center, which has been underway for more than a decade, to investigate the response and adaptations of various animal species to hypergravity.

This study on dogs was made possible through the development of a unique 7.9 m radius centrifuge designed to maintain relatively large sized animals under continuous hypergravity conditions without periodic service stoppages.

2. Procedures and Methods

Sixteen five-month old male Beagle dogs were employed; eight dogs were placed in individual swivel mounted cages on the centrifuge, while an equal number of controls were maintained in similar sized cages under normal gravity. The mean starting body weight ± S. E. of the centrifuged group was 7.35 ± 0.16 kg and 6.96 ± 0.22 kg for the control group. Dogs placed on the centrifuge were subjected to 1.5 *g* during the first week, 2.0 *g* during the second week, and 2.5 *g* for the remaining ten weeks of centrifugation. The floor dimensions of the cages (1.1 m × 0.74 m) permitted a moderate amount of physical movement. For heart rate and deep body temperature biotelemetry measurements, dogs were implanted with the sensor-transmitter units in the abdominal cavity one month prior to centrifugation. The equipment and procedure employed for biotelemetry measurements during centrifugation was basically the same as that previously used on rats [3]. All dogs were fed *ad libitum* daily; water was accessible at all times. A 12-hour light-dark lighting schedule was maintained throughout the study. At the end of the 12th week, all dogs were fasted overnight and bled from the femoral artery under sodium thiamlal (Surital) anesthesia and subsequently sacrificed with an overdose of pentobarbital. They were autopsied and individual organs and tissues weighed. Fore- and hindleg muscles and bones were dissected and weighed. Bones were also extracted with petroleum ether and diethyl ether and dried at 110 °C for fat-free dry weight measurements. Bone densiometric measurements were performed by photon absorptiometry [4] on the femur and humerus bones which were imbedded in paraffin wax and scanned at 5 mm intervals along the entire length of each bone. Blood analyses were performed by Solano Laboratories, Berkeley, California, using standard clinical methods.

3. Results

3.1. Growth

Centrifuged dogs showed a slower rate of growth than did the normal gravity controls (Fig. 1). Although the centrifuged group had an average starting body weight slightly higher than the control group, after 12 weeks of exposure the centrifuged group had an average body weight which was 0.62 kg lower than controls. In addition to the body weight changes, centrifuged dogs also showed significant decreases in the weights of their liver, spleen, heart, and abdominal fat.

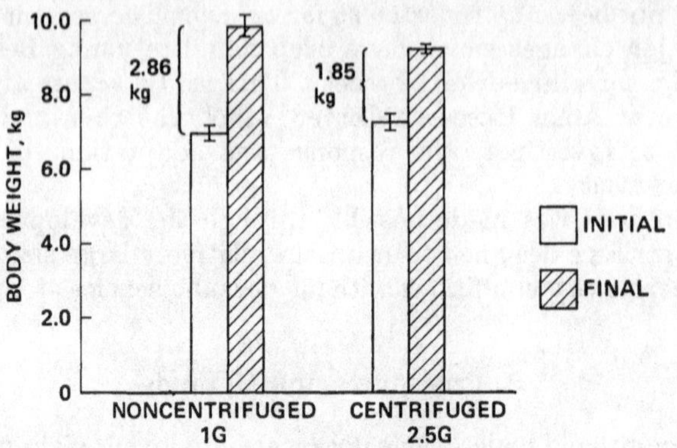

Fig. 1. Initial and final body weights of control and centrifuged dogs. Means ± S. E. of 8 dogs are shown.

3.2. Biotelemetry Measurements

Centrifugation induced an immediate and substantial decrease in deep body temperature of the young dogs (Fig. 2). Although the temperature was restored within the first 24 hours, significant changes in the amplitude and period length of the temperature rhythm persisted for several weeks (Fig. 3). A clear and dis-

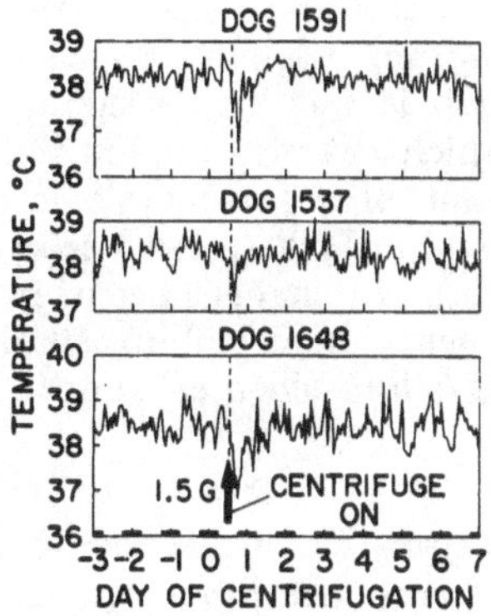

Fig. 2. Deep body temperature patterns in 3 young Beagle dogs prior to and during chronic centrifugation at 1.5 g. Twelve-hour on-off lighting schedule is shown along abscissa. Curves were drawn from hourly averages obtained from recordings taken every 10 minutes.

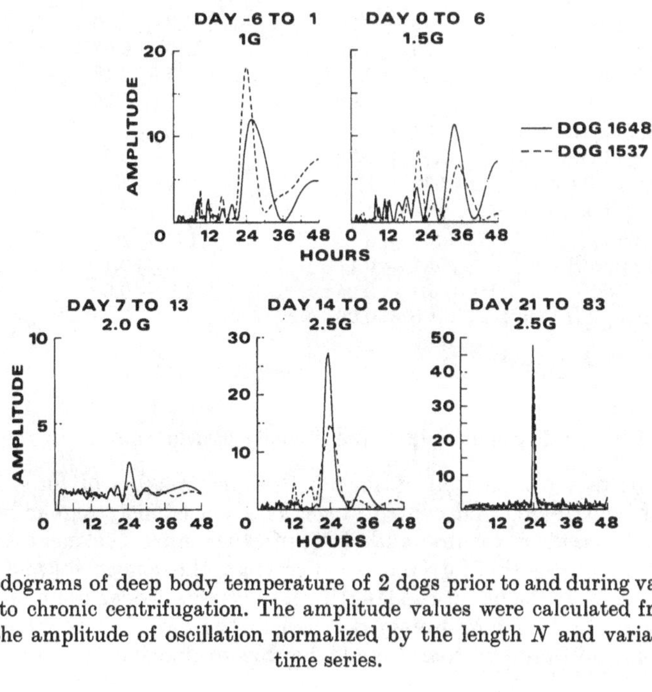

Fig. 3. Periodograms of deep body temperature of 2 dogs prior to and during various periods of exposure to chronic centrifugation. The amplitude values were calculated from: $NR^2/4\sigma^2$ where R is the amplitude of oscillation normalized by the length N and variance σ^2 of the time series.

cernible circadian rhythm was seen only after the second week of centrifugation. Heart rate patterns, which were also recorded at the same time, showed considerably more variation but generally responded in a similar manner to the temperature response.

3.3. Blood Analyses

Differences in selective serum constituents of centrifuged and control dogs are given in Table 1. A significantly lower protein concentration, due mainly to a decrease in albumin fraction was seen in the centrifuged dogs along with a lower total lipid concentration. Serum calcium was also lowered and although this decrease was relatively small it was highly significant ($P < 0.005$). Thyroxine (TT-4), the only hormone which was measured in this study, showed a significant increase. Table 2 shows some of the hematologic changes which were found. Significant changes were noted in PCV (packed cell volume), RBC (red blood cell count) and Hgb (hemoglobin). No changes occurred in MCV (mean corpuscular volume), MCH (mean corpuscular hemoglobin), MCHC (mean corpuscular hemoglobin concentration), WBC (white blood cells) or in the differential counts.

Table 1

Comparison of Serum Constituents

Values are means ± S. E. of 8 dogs with percentage changes given for only statistically significant differences between control and centrifuged dogs

Constituent	Noncentrifuged controls (8)	Centrifuged (8) 2.5 *g*	% Change
Total protein (g/100ml)	5.8 ± 0.06	5.3 ± 0.10	−8.0***
Albumin (g/100 ml)	3.29 ± 0.09	3.00 ± 0.08	−8.0*
Globulin (g/100 ml)	2.52 ± 0.11	2.35 ± 0.12	
A/G ratio	1.27 ± 0.08	1.27 ± 0.09	
Urea nitrogen (mg/100 ml)	13.4 ± 0.4	13.8 ± 0.9	
Creatinine (mg/100 ml)	0.71 ± 0.02	0.74 ± 0.03	
Lipids, total (mg/100 ml)	632 ± 18	551 ± 23	−12.8*
Cholesterol (mg/100 ml)	161 ± 5	150 ± 9	
Sodium (meq/liter)	147 ± 3	146 ± 1	
Potassium (meq/liter)	4.1 ± 0.1	4.3 ± 0.1	
Calcium (meq/liter)	5.1 ± 0.04	4.9 ± 0.05	−3.4***
TT-4 Thyroxine (μg/100 ml)	1.5 ± 0.06	1.9 ± 0.15	+24.0*

* $P < 0.05$, *** $P < 0.005$

3.4. Bone Weight and Density Measurements

Table 3 shows a comparison of the fat-free dry weights of fore- and hindleg bones of the centrifuged and control dogs. The consistently higher bone weights of the centrifuged dogs assume added significance since the mean body weight of this group (9.21 ± 0.17 kg) was smaller than the controls (9.83 ± 0.33 kg). No differences were found in the length and cross-sectional dimensions at midshaft of either the femur or humerus between the two groups which suggested that the weight differences found could be due to density differences. This was

Table 2
Hematologic Values
Means ±S. E. of 8 dogs are shown. Only statistically significant percentage changes are tabulated. For symbols (PCV etc.) see text

Analysis	Noncentrifuged controls (8)	Centrifuged (8) 2.5 *g*	% Change
PCV (%)	39.6 ± 0.8	34.4 ± 1.0	−13.0***
RBC ($10^6/mm^3$)	5.60 ± 0.13	4.88 ± 0.19	−12.8**
Hgb (g/100 ml)	14.0 ± 0.3	12.5 ± 0.4	−10.7**
MCV (μ^3)	71.2 ± 0.5	71.9 ± 0.3	
MCH ($\mu\mu$g)	25.1 ± 0.2	25.6 ± 0.2	
MCHC (%)	35.3 ± 0.2	35.5 ± 0.6	
WBC ($10^3/mm^3$)	8.9 ± 1.2	7.9 ± 1.0	
Differential (%)			
Polymorphs (Neutrophiles)	62.6 ± 2.6	67.2 ± 2.3	
Lymphocytes	26.4 ± 3.0	21.0 ± 2.0	
Bands	2.2 ± 1.0	2.4 ± 0.9	
Eosinophiles	0.7 ± 0.4	0.4 ± 0.2	
Monocytes	6.8 ± 1.5	7.0 ± 0.8	
Atypical lymphocytes	1.2 ± 0.4	1.6 ± 0.3	

** $P < 0.01$, ***$P < 0.005$

Table 3
Bone Weight Comparisons
Values are means ±S. E. of 8 dogs

Bone	Fat-free dry weight (grams) ± SE		%
	Noncentrifuged controls (8)	Centrifuged (8) 2.5 *g*	
Foreleg			
Humerus	14.11 ± 0.36	15.28 ± 0.31	+8.2*
Radius	6.15 ± 0.17	6.99 ± 0.16	+13.7***
Ulna	6.48 ± 0.20	7.93 ± 0.11	+22.3***
Scapula	7.98 ± 0.24	8.76 ± 0.26	+9.7
Hindleg			
Femur	15.51 ± 0.28	16.81 ± 0.29	+8.4**
Tibia	12.76 ± 0.29	13.68 ± 0.24	+7.2*
Fibula	1.31 ± 0.07	1.42 ± 0.05	+8.1

* $P < 0.05$, ** $P < 0.01$, *** $P < 0.005$

confirmed subsequently by density measurements of both the femur and humerus. Table 4 shows the bone mineral content distribution in g cm^{-1} axial length of bone at ten different equally spaced sites along the length of the humerus from scans of eight centrifuged and eight control dogs. Similar distributive changes in the distal portion of the femur were also observed. In contrast to the marked differences in the patterns of mineral content and in the total weight of the bones, no significant differences were found in the muscle weights of either the fore or hind limbs.

Table 4

Bone Density Distribution Patterns in Humerus by Photon Absorptiometry

Values are means of 8 dogs at each of the sites indicated in the figure. *P* value indicates level of significance between control and centrifuged values

	ASH WEIGHT (g/cm)		
SCAN SITE	CONTROL (8)	CENTRIFUGED (8)	P VALUE
10	.974	.935	
9	1.049	1.184	<.05
8	.588	.643	
7	.594	.649	
6	.555	.614	<.005
5	.536	.596	<.0005
4	.517	.580	<.005
3	.545	.613	<.005
2	.522	.594	<.005
1	.966	.930	
BONE WEIGHT (g)	14.11	15.28	<.05

4. Discussion

The slower rate of growth shown by the centrifuged dogs was anticipated since most animals which have been subjected to chronic centrifugation show the same response. The relatively modest degree of growth suppression effected in this study, however, indicates that the dogs could tolerate a higher *g*-load without too much difficulty.

The transient hypothermic response and alteration in the normal diurnal temperature rhythm seen in the centrifuged dogs is quite similar to the response of rats to centrifugation [3]. Temperature measurements, however, recorded in dogs older than one year show considerably less temperature fall upon centrifugation even though the initial *g*-load may be considerably higher, i.e. 2.5 *g*. The ability to thermoregulate during an imposed gravitational stress is clearly dependent on the age of the animal, the younger dogs being less capable of maintaining their normal body temperature than the older dogs. Results of this study indicate that the deep body temperature rhythms may be one of the more sensitive indices by which to assess the overall state of physiologic adaptation of animals to hypergravity. In order to confirm this, correlative measurements of body weight and food consumption changes will have to be made. It should be noted that the hypothermia experienced by the dogs when first subjected to centrifugation at 1.5 *g* was not detected when the dogs were subsequently exposed to 2.0 *g* or 2.5 *g* indicating habituation to centrifugation.

Decreases in plasma protein concentrations found in the young dogs of this study are also found in older dogs after long-term centrifugation. The basis for this decrease is not known. The small but, nevertheless, highly significant decrease in serum calcium levels also appears to be a general response to hypergravity and is seen in both young and old dogs as well as in chronically centrifuged rats [5]. Determinations of circulating hormone levels which are known to affect calcium levels in the blood, particularly calcitonin and parathyroid hormone, may help resolve this problem. The increase in thyroxin levels in the hypergravity adapted dogs is compatible with an increase in metabolic rate. Although no metabolic rate measurements on chronically centrifuged dogs have been made as yet, there is every reason to expect that the dogs will show an increased metabolism. Hypergravity-adapted rats, for example, have been shown to have as much as a 25% higher resting metabolic rate than normal gravity controls [6].

The reason for the observed decrease in red blood cells of centrifuged dogs is not known. The cells which are present are normal in all respects as compared with the cells of the control dogs. It is planned that future studies on hypergravity exposure effects will include investigations on erythropoietic function.

This study is believed to be the first which demonstrates an increase in bone density in animals subjected to prolonged hypergravity. Although increases in either the mass or density of the load bearing leg bones of centrifuged animals have been predicted, experimental results have indicated either no change or actual decreases in bone mass [7, 8]. Fosse [9] who investigated bone density changes in chronically centrifuged mice found no changes in male mice and decreases in female mice. The positive effects on bone density found in dogs may reflect a critical dependence on the body size of the animal exposed to hypergravity. A current study underway, in which three-month old dogs are being centrifuged for six months with periodic *in vivo* bone density measurements being made, should further clarify the density changes obtained in this study.

Acknowledgments

The author thanks L. Chan, W. T. Platt, B. Daligcon and R. W. Johnson for their technical assistance, and G. L. Fisher and Dr. M. Goldman (Radiobiology Laboratory, University of California, Davis) for performing the photon absorptiometry measurements and analyses on bone.

References

[1] A. H. Smith and R. R. Burton, in: Gravity and the Organism, University of Chicago Press, Chicago 1971 (p. 371).
[2] C. C. Wunder, in: Gravity and the Organism, University of Chicago Press, Chicago 1971 (p. 389).
[3] J. Oyama, W. T. Platt and V. B. Holland, Am. J. Physiol. **221**, 1271 (1971).
[4] J. A. Sorenson and J. R. Cameron, J. Bone and Joint Surgery **49 A**, 481 (1967).
[5] J. Oyama and B. Zeitman, Am. J. Physiol. **213**, 1305 (1967).
[6] J. Oyama and L. Chan, Fed. Proc. **32**, 392 (1973).
[7] J. P. Jankovich, NASA Contractor Report 1823 (1971).
[8] L. Keil, Doctoral Dissertation, University of California, Davis (1973).
[9] G. Fosse, Growth **35**, 33 (1971).

Life Sciences and Space Research XIII — Akademie-Verlag, Berlin 1975

BODY COMPOSITION CHANGES WHICH ARE PROPORTIONAL TO CHRONIC ACCELERATION

G. C. PITTS[a], L. S. BULL[b] and J. OYAMA[c]

[a] Physiology Department, University of Virginia, Charlottesville, Va, USA
[b] Department of Dairy Science, University of Maryland, College Park, Md, USA
[c] Biomedical Research Division, NASA Ames Research Center, Moffett Field, Calif., USA

Chronic centrifugation (2.7 *g*, 3.18 *g*, 4.15 *g*) of the rat (both of the adults and weanlings) results in an apparent dissociation of regulation of the fat-free compartment from regulation of energy balance which can be demonstrated by body composition analyses. The levels of energy balance before and during centrifugation were altered by special diets (high-protein, high-fat, chow) and by partial inanition. Changes in both body fat and the fat-free body compartment were followed in both weanlings and young adults: The principal effector mechanism involved in the reduction of the fat-free compartment during the first few days of centrifugation was feeding behavior. This loss of mass was independent of energy balance or nitrogen balance (within limits) and determined apparently solely by the change in acceleration. We have concluded that the change in *g*-force reduced the size of the fat-free compartment in a direct manner over which the body had no regulatory control. After the new level was established, mechanisms regulating growth resumed function.

A full paper is scheduled for March 1975 in the *American Journal of Physiology.*

Life Sciences and Space Research XIII — Akademie-Verlag, Berlin 1975

GRAVITATIONAL EFFECTS ON BODY COMPOSITION IN BIRDS

A. H. Smith, O. Sanchez P.* and R. R. Burton**

Chronic Acceleration Research Laboratory, Department of Animal Physiology, University of California, Davis, Calif., USA

Gallinaceous birds, presenting a wide range of body size, were adapted physiologically to hyperdynamic environments, provided by chronic centrifugation. Chemical composition was measured directly on prepared carcasses, which were anatomically comparable, and more amenable to analysis than the intact body. Body mass and body fat decreased arithmetically with increasing field strength and also with increasing body mass. Water content of lean tissue increased in hyperdynamic environments, but irrespectively of body size.

The interaction of size and gravity is an important relationship which was first discussed, as the "Principle of Similitude", by Galileo [1], and it also appears to be a determinant in responses of animals to chronic acceleration [2]. Reductions of body mass and body fat in chronically accelerated animals have been reported by several investigators, and with a variety of experimental animals, and these observations have been reviewed in previous reports [2, 3]. Rationalization of such gravitational effects has been complicated by the heterogeneity of experimental animals, so that size and species influences are not readily differentiated. Consequently, body size and composition measurements were made on an acceleration-adapted series of similar animals, gallinaceous birds, with a size range of 0.1 to 6 kg body mass, to clarify the interaction of the acceleration field and body size upon body composition.

Experimental animals were Japanese Quail (*Coturnix coturnix japonica*) and two breeds of domestic fowl (*Gallus gallus domesticus*). One of the latter was a White Leghorn (SCWL) strain selected for acceleration tolerance [4] and the other was a commercial meat-producing hybrid, "Arbor Acre". Acceleration fields were provided by animal centrifuges which have been described elsewhere [5, 6]. Operation of these machines was quasi-continuous, being interrupted only briefly, daily, for feeding, cleaning and observation of animals. Observations on other animals exposed to increased acceleration fields for varying daily periods [7] indicate that there is a time-intensity summation for vegetative phenomena such as reported herein, so the brief daily interruptions were not considered significant to the results.

The carcass, which approximates the edible portion, was prepared by removal

---

* Present address: Ministry of Agriculture, Bogota, Colombia, S. A.

** Present address: Biodynamic Branch, USAF School of Aerospace Medicine, Brooks AFB, Texas, USA.

of the skin, shanks (the leg below the hock joint), head and shed blood. After some measurements of visceral size (not reported herein) the entire carcass was comminuted in a commercial meat grinder, and an aliquot removed for chemical analysis. This sample was dried to a constant weight at 105 °C, and the dry residue extracted with petroleum ether. The mean values for body mass and the proximal analysis of each group are summarized in Table 1. Among the three groups of

Table 1
Proximal Analysis

The results of chemical analysis of carcass samples are summarized by groups as the mean ± standard deviation

Group	Field		Body Mass	Carcass Composition		
				Water	Fat	Lean
	(g)	(n)		(%)	(%)	(%)
Quail	1	(10)	116.1 ± 7.5 g	65.0 ± 3.5	10.90 ± 4.15	24.12 ± 1.86
	2	(5)	122.6 ± 6.7	66.0 ± 4.6	10.64 ± 4.53	23.36 ± 1.30
	2.5	(6)	106.7 ± 2.7	70.2 ± 2.4	5.60 ± 2.30	24.35 ± 0.79
	4	(2)	99.6 ± 0.7	71.5 ± 0.27	1.21 ± 0.05	27.26 ± 0.22
Leghorn Chicken						
P ♂	1	(6)	1.33 ± 0.07 kg	65.9 ± 3.0	6.67 ± 2.59	27.45 ± 0.79
P ♀	1	(6)	1.07 ± 0.29	64.4 ± 5.3	9.38 ± 5.41	26.20 ± 2.19
P ♂	3	(6)	0.94 ± 0.10	70.6 ± 1.0	1.39 ± 0.41	28.02 ± 1.09
P ♀	3	(6)	0.71 ± 0.03	70.9 ± 1.1	1.71 ± 0.69	27.40 ± 0.59
T	1	(5)	2.22 ± 0.21	65.4 ± 2.7	8.26 ± 4.01	26.36 ± 1.38
	1.5	(6)	2.00 ± 0.20	68.3 ± 1.2	5.32 ± 1.41	26.32 ± 0.69
	2	(5)	1.88 ± 0.17	69.4 ± 1.1	4.34 ± 1.57	26.28 ± 0.82
Z	1	(13)	2.29 ± 0.19	54.2 ± 3.7	22.94 ± 4.70	22.88 ± 1.75
	1.5	(5)	1.94 ± 0.36	62.2 ± 5.0	13.34 ± 6.88	24.44 ± 2.02
	2	(7)	1.90 ± 0.15	69.2 ± 2.3	3.40 ± 1.50	27.43 ± 1.04
	2.5	(10)	1.87 ± 0.17	70.0 ± 1.1	2.61 ± 1.08	27.38 ± 0.68
Arbor Acre	1	(14)	5.39 ± 0.54	61.6 ± 4.8	14.94 ± 5.53	23.71 ± 1.47
	2	(5)	4.02 ± 0.31	72.1 ± 3.4	2.66 ± 1.38	25.12 ± 2.33

n = number of observations

Leghorn chickens, designated P, T and Z, there are some obvious differences in body size and carcass composition. Group P was composed of young animals, 4—5 months of age, whereas the others were mature, being 1—2 years of age at time of sacrifice. This is the basis for the lesser body size and carcass fat concentration of the P group. Group T was used in a series of feed balance studies, to determine maintenance requirements [8]. Frequently this involved feed intakes less than *ad libitum*, which accounts for their low fat concentration. The loss of fat per unit increase in field strength was similar for the mature T and the younger P birds, and less than for the more adipose Z birds.

All of the parameters summarized in Table 1 appear to change more or less arithmetically with increasing field strength. The gravitational effect was esti-

Table 2

Influence of the Acceleration Field Strength Upon Body Size and Composition

Body mass, carcass water and carcass fat concentrations all appear to vary arithmetically with field strength g, and they can be described by the equation $y_g = y_0 + kg$. Regressions of group mean data upon acceleration field strength are summarized below for several parameters by the appropriate constants for the general equation and by the correlation coefficient r.

		Body Mass			Carcass Water (%)			Carcass Fat (%)		
	(n)	M_0	$-k_m$	r	W_0	k_w	r	F_0	$-k_f$	r
Leghorn Chickens										
Group P (1—3 g)	(4)	1.39 kg	—0.187	—0.837	62.34	2.80	0.982*	11.26	—3.24	—0.958*
Group T (1—2 g)	(3)	2.54 kg	—0.340	—0.986	61.69	3.98	0.969	11.85	—3.92	—0.961
Groups P and T	(7)	2.10 kg	—0.364	—0.565	62.96	2.72	0.944**	11.07	—3.21	—0.955***
Group Z (1—2.5 g)	(4)	2.46 kg	—0.260	—0.859	44.84	10.89	0.955*	35.40	—14.19	—0.956
Quail (1—4 g)	(4)	127.2 g	—6.93	—0.880	62.67	2.32	0.914	15.35	—3.48	—0.942
Arbor Acre (1—2 g)	(2)	6.76 kg	—1.37	—	51.12	10.47	—	27.22	—12.28	—

Random probability: * <0.05; ** <0.01; *** <0.001

n = number of observations

mated by fitting the mean data for various fields to the equation

$$y_g = y_0 + kg$$

where y_g is the magnitude of the parameter in animals adapted to an acceleration field of g strength; y_0 is the intercept, the magnitude indicated for weightlessness (the implied continuity of effects of hyper- and hypodynamic environments has been considered in [9]); k is the proportionality coefficient, which is positive for water concentration, but negative for body mass and fat concentration.

The results of this mathematical analysis are summarized in Table 2, as the constants for the preceding equation, and the correlation coefficient r. Although the correlation coefficients generally are large, only a few of the regressions have statistical significance, a consequence of the small samples being compared.

A convenient means of comparing the gravitational effect among groups of varying body size is by the ratio of proportionality coefficient k to the intercept y_0, expressed as a percentage. Implicit in this ratio, designated I, is the inhibitory effect of earth gravity upon the phenomenon:

$$\% I = -k/y_0 \times 100.$$

The gravitational inhibition of potential mature body mass is compared with body size in Fig. 1. For this relationship, the P Leghorns (4—5 months of age)

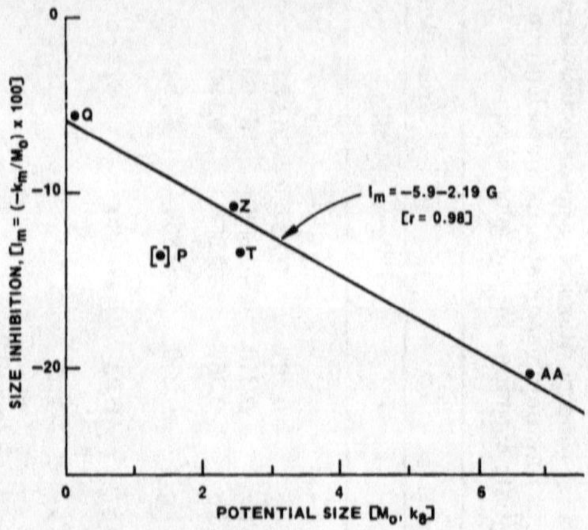

Fig. 1. Potential body size and gravitational size limitation.

The relative severity of increasing acceleration fields upon mature body size can be compared by the ratio: $-k_m/M_0$, expressed as %. This ratio, designated I_m, implies the relative reduction in potential mature body mass imposed by earth-gravity. The line indicates the relationship, omitting P because of their young age, which has a random probability of 0.02. For all groups, the equation is: $I_m = -7.5 - 2.0M_0$ [$r = -0.91$; $p = 0.04$].

were omitted, since immature animals are more severely stressed by chronic acceleration than are mature animals [4]. Larger animals are more affected by acceleration, the size-inhibition I in animals of 4 kg body mass being double that in those of 1 kg.

Gravitational inhibition of fat deposition I_f is compared with the potential body fat concentration for weightlessness F_0 in Fig. 2. Although some relationship does appear to exist, it is neither marked, nor statistically significant. Potential body size for weightlessness M_0 has a much greater influence over the gravitation inhibition of fat I_f, as indicated in Fig. 3. Although the exponential kinetics ($I_f = aM_0^b$) provide a better fit to the data than that of arithmetic kinetics ($I_f = a + bM_0$), it is difficult to conclude the nature of the relationship. The

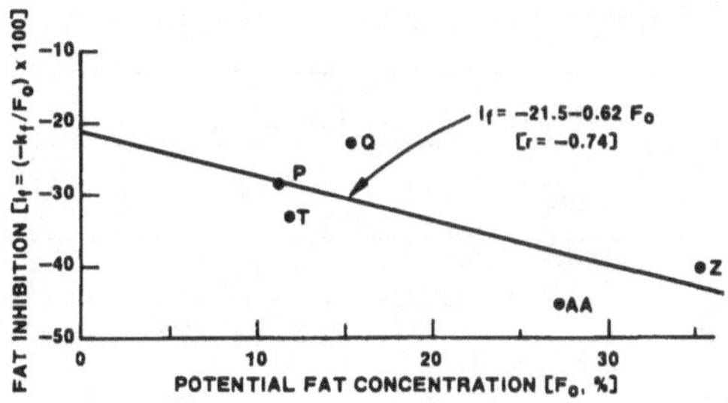

Fig. 2. Potential fat content and gravitational limitation of adiposity.

The influence of increasing acceleration fields in limiting body fat I_f is compared with the potential fat content F_0. This relationship, indicated by the line, has a random probability of 0.15 at this sample size and lacks statistical significance.

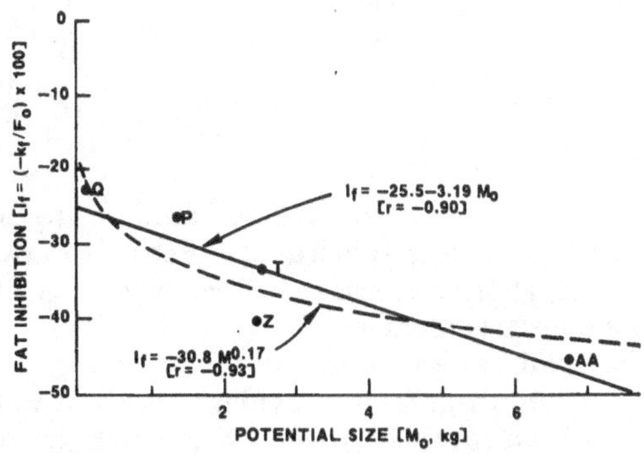

Fig. 3. Potential body size and gravitational limitation of adiposity.

The influence of increasing acceleration field strenght in limiting adiposity I_f appears to be closely related to potential mature body mass M_0. Two relationships are shown, the parabolic kinetics ($p = -0.02$) fitting the data better than the arithmetic kinetics ($p = -0.04$).

numbers of groups under comparison are few, and the physiological basis for the relationship is not apparent. It should be noted that no marked nor significant relationship is apparent between the potential fat concentration (F_0, %) and the potential mature body mass (M_0, kg):

$$F_0 = 14.9 + 2M_0 \quad \text{(for 5 groups, } r = 0.473\text{)}.$$

The general inverse relationship between body fat and body water has been recognized for some time [10]. According to this principle, the water content of the lean body tissue is uniform, and no significant body water is contributed by adipose tissue. The arithmetic relationships between carcass fat (F, %) and carcass

Table 3

Carcass Fat and Water Relationships

There is a general inverse relationship between water ($W_{\%}$) and fat ($F_{\%}$) concentrations as a result of the markedly different hydrations of lean and adipose tissues: $W_{\%} = W_{\mathrm{L}} - b\,(F_{\%})$ where W_{L} is the water content of lean tissue (i.e. $W_{\%}$ when $F_{\%} = 0$; b is the proportionality coefficient. The water-fat relationships at several acceleration field strengths are summarized as the constants for the equation and the correlation coefficient r.

		Field (g)	Water-Fat Equation (n)	W_{L}	b	r	Mean Fat Concentration % ± SD
Immature SCWL	P	1	(12)	72.3	−0.89	−0.920***	8.02 ± 4.29
($M_0 = 1.39$ kg)	P	3	(12)	72.3	−0.99	−0.897***	1.55 ± 0.56
Mature SCWL	T & Z	1	(40)	71.0	−0.72	−0.981***	15.18 ± 7.79
($M_0 = 2.50$ kg)	T & Z	1.5	(11)	72.1	−0.74	−0.991***	18.97 ± 23.6
	T & Z	2	(12)	72.8	−0.94	−0.800***	8.29 ± 7.23
	Z	2.5	(10)	72.1	−0.80	−0.796**	2.61 ± 0.18
Quail		1	(10)	73.3	−0.76	−0.901***	10.90 ± 4.15
($M_0 = 0.127$ kg)		2.25	(11)	75.6	−0.92	−0.967***	7.89 ± 4.22
Arbor Acre		1	(14)	73.8	−0.82	−0.942***	14.94 ± 5.53
($M_0 = 6.76$ kg)		2	(5)	77.2	−1.93	−0.790	2.66 ± 1.38

Random probability: * <0.05; ** <0.01; *** <0.001; n = no. of observations

water (W, %) were compared for the various animal groups and acceleration fields, and the results are summarized in Table 3. For this analysis some additional birds were added to the Leghorn controls to increase the range of body composition. These were some previously centrifuged birds of the T group, after five months at earth-gravity, that had re-established the water-fat relationships typical of controls. From the data in Table 3, it is evident that the water content of lean tissue W_{L} increases with an increasing acceleration field in mature animals, but apparently not in immature ones (e.g. group P). Body size does not appear to exercise any influence upon this gravitational effect, the acceleration-increased hydration of Quail lean tissue being as great as in the larger Leghorn chickens. It is considered likely that the factors (largely hydrostatic pressures and vascular distensibility) which lead to increased plasma volume in chronically centrifuged animals [11] also enhance tissue water content. Such a mechanism should provide a proportionality between lean tissue water concentration and acceleration field strength, but this is not apparent in the data presented (e.g. mature Leghorns, Table 3). In centrifuged animals there is a progressive increase in the proportionality coefficient, b, with increasing field strength which is difficult to rationalize with the current concepts concerning tissue hydration. According to the Pace Principle [10], that tissue water represents the relative presence of lean tissue which has a constant hydration, the two coefficients, W_{L} and b, should be numerically equivalent. The numerically greater values for b in the mature animals, which is enhanced by chronic acceleration, implies that within these treatments adipose tissue not only dilutes lean tissue, but also dehydrates it. Alternatively, the inconsistencies with the data presented herein may be technical, arising from differences in preparation of tissues (an exsanguinated carcass *vice* intact body) and the very narrow range of fat content in the centrifuged animals.

Acknowledgment

This study was supported under National Aeronautics and Space Administration grant NGR 05-004-008.

References

[1] G. GALILEO, Dialogues Concerning Two Sciences, 1638. English translation by H. Crew and A. DeSalvio, Macmillan, New York, 1914.
[2] A. H. SMITH, Life Science and Space Research XI, 209 (1973).
[3] A. H. SMITH, D. F. RAHLMANN, A. M. KODAMA and N. PACE, Life Sciences and Space Research XII, 129 (1974).
[4] R. R. BURTON and A. H. SMITH, Aerospace Med. **36**, 39 (1965).
[5] A. H. SMITH and C. F. KELLY, Naval Research Rev. **18(11)**, 1 (1965).
[6] C. F. KELLY, A. H. SMITH and C. M. WINGET, J. Appl. Physiol. **15**, 753 (1960).
[7] R. R. BURTON and A. H. SMITH, Am. J. Physiol. **222**, 1505 (1972).
[8] A. H. SMITH, R. R. BURTON and C. F. KELLY, J. Nutr. **101**, 13 (1971).
[9] A. H. SMITH and R. R. BURTON, in: Gravity and The Organism, Univ. Chicago Press, Chicago 1971.
[10] N. PACE and E. N. RATHBUN, J. Biol. Chem. **158**, 685 (1945).
[11] R. R. BURTON and A. H. SMITH, Space Life Sci. **1**, 503 (1969).

Life Sciences and Space Research XIII — Akademie-Verlag, Berlin 1975

THE INFLUENCE OF VARIABLE GRAVITATIONAL FIELDS ON THE EMBRYONIC DEVELOPMENT OF SOME ECAUDATE AMPHIBIANS

V. V. Popov, L. R. Palmbakh and E. V. Kuznetsov

USSR, Academy of Sciences, Moscow, USSR

The fertilized eggs of the frog *Rana temporaria* immediately after fertilization were rotated on a clinostat in the vertical plane at the rate of 0.66 rotations per minute with a radius of 20 mm for 2.5 hours. A wide spectrum of developmental anomalies was found (33% in the experiment, 14% in the control) which, in the authors' opinion, result from abnormalities in the cortical reaction of symmetrization. These abnormalities manifest themselves in the irregular distribution of cortical pigment and in eccentric division lines.

The effect of gravitation on the early development of amphibians was clearly demonstrated at the end of the last century in studies by Pflüger [1] and Born [2]. Further investigations by Penners and Schleip [3], Pasteels [4, 5], Ancel and Vintemberger [6] partly explained the mechanism involved. The major method used in these studies consisted in changing the natural orientation of the fertilized egg in the earth's gravitational field. It was effected by turning the egg with the vegetative pole uppermost or orienting the egg so that the animal-vegetative axis and the vertical made a certain angle. These authors concluded that gravitation is a powerful polarizing environmental factor which affects the formation and determination of bilateral symmetry of a fertilized amphibian egg at an early stage of its development. Symmetrization of a fertilized egg, occupying the period from fertilization to the first cleavage and leading to formation of bilateral symmetry, is a key moment in the embryonic development of the organism.

The great number of embryonic abnormalities arising if eggs are turned or centrifuged in the period of symmetrization [7, 8] show the high sensitivity of fertilized eggs during this period to imposed and powerful (sometimes non-specific) forces. Such a disturbing effect of the environment on a fertilized egg is not observed, in fact, under natural conditions. The fertilized egg immediately starts turning over with its heavier part down and therefore its animal-vegetative axis is oriented strictly vertically and along the gravitational vector. Thus, developmental anomalies arising from turning, and correlated with changes in the gradient of yolk globules (due to movement downwards of the heavy yolk cap of the vegetative hemisphere) are not frequent under natural conditions. As for egg centrifugation, the use of a deliberately greater centrifugal force than the earth's gravitational field hampers interpretation of the specific action of natural gravitation.

Of particular interest in our opinion, for elucidation of the specific action of the earth's gravitational field on the early development of amphibians, are investigations with the clinostat in the vertical plane. Apparently, biological effects of gravitation are due not so much to its magnitude as to its constant and strict orientation. The use of clinostat rotation for elucidation of gravitation effects on the early development of amphibians can yield very interesting results [9].

Our work is aimed at studying some peculiarities in the development of eggs of the frog *Rana temporaria* in which symmetrization took place at a minimal oriented gravitation. Fertilized eggs were rotated in a vertical plane at the rate of 0.66 rev min^{-1} at a radius of 20 mm which amounted to 9.4×10^{-6} g.

Two series of experiments were performed on eggs from two mature females of *Rana temporaria*. The method of hypophysial injection was used to obtain mature hard-roe which was quickly and accurately fixed in a natural way with the help of a glass stick to the dry surface of a glass cylinder of radius 20 mm. The eggs were distributed on the cylinder surface in such a way that they lay freely in one layer. They were fertilized by a water suspension of sperm applied to the cylinder for 3 minutes, while the cylinder was slowly rotated by hand. Immediately afterwards the cylinder was installed on a horizontal axis of the clinostat and put into an aquarium with fresh water. In this tank the eggs were rotated for 2.5 hours at 20 °C till the first division line appeared. The control eggs were stuck to the bottom of a glass crystallizor.

After the experimental treatment the eggs (with much swollen coats) were carefully separated from the surface of the cylinder and removed into the crystallizor for further study. The material was fixed in Bouin's fluid after 2.5, 24, 48 and 72 hours and on the 5th day after the fertilization. The control material was fixed at the same intervals. Altogether 1009 egg were analysed in the experiment, and 1648 in the control. The percentage of fertilized eggs that entered division was on the average about 93%.

The study allowed detection of a number of characteristic anomalies at different stages of development. At the early stages of development in about one third of the eggs the first division line lay across the animal-vegetative axis. We also noted that pigment condensed at the animal pole of eggs rotated on the clinostat. Another anomaly consisted of migration of pigment deep into the eggs so that the cortical layer of the egg acquired a grey colour in the animal region. In other cases the division of the pigment into lines was observed. It should be pointed out that these frequent types of pigmental anomalies were often combined. As a result, the number of pigmental anomalies in the experiment was 17%, and in the control 4.7%. This difference is statistically significant.

Observations on the experimental material after 24 hours revealed a great number of division abnormalities which were combined with the abnormalities in pigmentation. Later, at the gastrula stage, different defects in gastrulation were found: these included slow rate of epiboly and invagination defects (asymmetric formation of dorsal and lateral lips of blastopore). Some eggs had ectoderms of very peculiar form; some embryos had cluster-like forms of ectoderm, others had ectoderm in the form of thorns.

At the later stage of neurula the developmental defects included weak expression of the neural tube, with partial or complete closure, the absence of stretching of the embryo in the cranio-caudal direction (the embryos had a rounded form) and remains of yolk. Some embryos had small local defects: depending on their location the yolk or mesoderm lay open.

The difference between the percentages of embryos with specific anomalies in the experiment and in the control is statistically significant.

In embryos fixed 72 hours after treatment and on the 5th day of the experiment anomalies were correlated with defects arising during formation of the axis complex. The most frequent were monstrous embryos with abdominal growths, embryos with incompletely ground in plugs, micro- and macrocephalies, embryos with a cauda split into two, and double monsters with differing degree of union at the head or the tail. As a rule, microcephaly was combined with ectoderm growth on the ventral part of an embryo. It should be noted that ectodermal growths in embryos at all the stages investigated were much pigmented. Less frequently experimental embryos had local regions of epidermis with poor pigmentation. It should be also emphasized that on the 5th day of the experiment the experimental

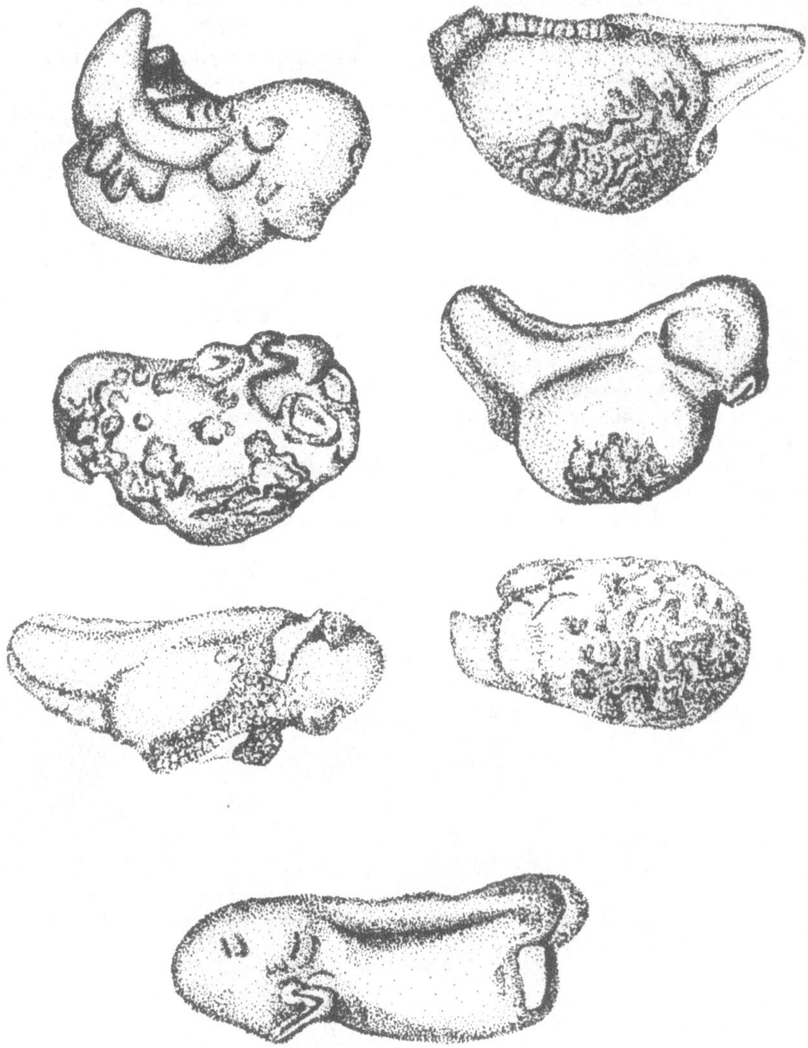

Fig. 1. Developmental anomalies on the 5th day after experimental treatment.

embryos were smaller than the control ones. The number of all developmental anomalies on the 5th day was 33% in the experiment and 14% in the control.

These anomalies indicate a specific effect of gravitation on the process of symmetrization. Anomalies in the pigment distribution and abnormally situated first division line seem to indicate some disorder in the cortical reaction to symmetrization at the period of determination of bilateral symmetry of developing eggs. Anomalies at the next stages of development (disturbances in cleavage or gastrulation, anomalies in the formation of the axis complex, local ectodermal growths etc.) are entirely determined by effects produced at this important period. Our results agree with those of Tremor and Souza [9].

The studies of developing amphibian eggs in a variable gravitational field indicate an important role of gravitation in the early development of vertebrates with a holoblastic type of development. The earth's gravitational field is evidently an important environmental factor that stabilizes polarization of certain physico-chemical process in developing eggs, which in turn determine morphogenesis.

It will be important in the future to follow the development of clinostated animals till the stage of late tadpole or young frog: this will enable us to find out whether the defects reported above can affect the process of late ontogenesis or whether these defects could be repaired in the course of development.

References

[1] E. Pflüger, Pfl. Arch. **32**, 1 (1883).
[2] G. Born, Arch. mikr. Anat. **24**, 475 (1885).
[3] A. Penners and W. Schleip, Z. Wiss. Zool. **131**, 1 (1928).
[4] I. Pasteels, Arch. Biol. **50**, 292 (1939).
[5] I. Pasteels, Bull. Soc. Zool. Fr. **76**, 231 (1953).
[6] P. Ancel and P. Vintemberger, Bull. Biol., Suppl., **31**, 1 (1948).
[7] V. L. Kas'yanov, Dokl. Akad. Nauk USSR **180**, 1008 (1968).
[8] R. S. Young, P. H. Deal, K. A. Souza and O. Whitfield, Exper. Cell. Res. **59**, 267 (1970).
[9] I. W. Tremor and K. A. Souza, Space Life Sci. **3**, 179 (1972).

Life Sciences and Space Research XIII — Akademie-Verlag, Berlin 1975

OTOLITH FUNCTIONS IN WEIGHTLESSNESS

S. J. Gerathewohl

Federal Aviation Administration, Office of Aviation Medicine, Washington, D. C., USA

The role of the vestibular organ in the exploration of space has been studied extensively during the past two decades. Many investigators have shown that some persons experience ill effects during the transition from the normal gravity to subgravity or weightlessness. Such adverse reactions can be related to a variety of sensory and somatic changes within the body systems; but it appears that the two major components of the unusual force field—namely, the absence of gravitational stimulation of the otolith organs and the occasional stimulation of the semicircular canals by head and body movements—bring about the motion sickness type reactions. Experiments in parabolic flights and in spacecraft revealed that the statolith organs respond to changes of acceleration during zero G. After an initial period of increased activity during the transition from 1 G to zero G, the number of nerve impulses from the otoliths is drastically decreased and later becomes steady on a somewhat lower than normal level of the discharge rate. The various theories concerning otolith responses in weightlessness are discussed and validated against the actual findings on astronauts and cosmonauts during spaceflight experiments and missions.

1. Introduction

The role of the vestibular organs in the exploration of space has been studied extensively during the past 25 years. After the alarming statement in 1949 that space travelers may become incapacitated by disturbances of the gravireceptors during zero G [1], the otolith functions were studied theoretically as well as experimentally on animals and human subjects in the laboratory, airplanes, rockets, and spacecraft [2]. Before summarizing the results which are now available on this subject, a few assumptions which provide a basis for this discussion must be made. First, it is assumed that the principle of d'Alembert can be applied to describe the gravito-inertial force relationship of the weightless state. This concept permits the conclusion that the sum of all forces acting on a body is equal to zero under all conditions of rest or motion [3]. Secondly, it is assumed that gravitational, accelerative and inertial forces have identical physiological and biomedical effects [4]. Thirdly, it is assumed that the states of subgravity or weightlessness obtained through free fall, parabolic flight, or in orbit are substantially identical as to their physical, physiological, and biological effects [5]. Finally, it is assumed that the otolithic organs are normally stimulated by changes of weight and linear accelerations. The threshold value of otolith stimulation in man has been given as 3×10^{-4} g [6].

2. Experiments with Animals during Weightless Conditions

Early experiments conducted by Henry et al. [7], von Beckh [8], Schock [9], and King [10], using normal versus vestibularly damaged animals during short periods of weightlessness in aircraft and rockets, showed that the normal specimens moved uncontrolled during zero *G* and showed symptoms of disorientation and panic; whereas, the injured or labyrinthectomized animals were more controlled and capable to cope with the weightless situation [7—11]. The labyrinthine posture reflex or righting reflex of the normal cat was delayed or failed completely after a period of about 20 seconds of virtual weightlessness [12]. Investigations of reflex behavior and oculomotor reactions during increased accelerations and zero *G* conducted by USSR scientists also showed that a close relationship existed between otolith stimulation through linear accelerations and the tonic reflexes originating in the vestibular organ [13]. The electrical activity at the vestibular ganglion in unanesthetized cats during 1.2 seconds of free fall was recorded by Fiorica et al. [14]. The firing rate changed significantly in this condition; and this change usually triggers the labyrinthine righting reflex. Goldfish lost their orientation in weightlessness. Blinded, they committed "diving loops" [15].

From these experiments, it was concluded that the otoliths, losing weight, do not function properly as gravireceptors and cause changes of the normal reflex pattern, loss of motor control, and spatial disorientation in experimental animals. This seems to be particularly true for "three dimensional animals", such as fish, turtles, birds, cats, monkeys, and other species that move with ease and skill in all three coordinates of space. It must be pointed out, however, that weightlessness did not cause a functional exclusion of the otoliths, but produced an unusual excitation of the gravito-inertial analyzer. Evidently, weightlessness and loss of the labyrinth are not the same; but the mechanisms involved in the deafferentiation process of the otoliths remained unclear.

The classical experiments on the recording of nervous signals from the otolith organs were made by Adrian in 1943 on the cat, and by Lowenstein and Roberts in 1949 on the thornback ray (*Raja clavata*) [16, 17]. Recently, Gualtierotti (in 1964—1972) has recorded the electric discharge rate from individual fibers of the vestibular nerve of the frog [18]. During the recording of action potentials in the macula utriculi, it was found that the slightest head movement of the specimen caused variations of the electrical impulse frequency. These changes were closely associated with a change in the magnitude of the gravitational or acceleration vector acting on the statoconia. By means of tilting experiments and in parabolic flights, it was demonstrated that the otolith organ at single unit level is subjected to a profound change in its activity as a result of going from 1 *G* to zero *G*. There was an immediate increase of spontaneous firing at the start of the zero-*G* period, a sudden suppression of response after about 10 seconds of weightlessness, an initial large response to post-weightlessness acceleration during pull-out from the dive, and the restoration of the normal discharge rate after return to the 1 *G* condition.

During $6^1/_2$ days in orbital flight and less than 10^{-3} *G*, there was an alteration of the firing rate when the animal was at rest, first decreasing and then increasing at an approximate 24-hour period [19]. The impulse frequency returned to normal after approximately 90 hours of space flight. During the periodic centrifugation at an acceleration of 0.6 *g*, the impulse frequency increased slightly with increasing

G-forces, but decreased again at the end of each acceleration period. After five days in orbit, the responses both from the resting and from the repeatedly accelerated animal were practically back to normal.

The most widely accepted concept concerning otolith function today is that the reception in the statoconia is activated by the displacement of the otolithic membrane relative to the receptor cells [20]. Any displacement of the otoliths, whether perpendicular to the macula or through shearing motion, acts on the cilia of the sensory cells. It has to be considered that under 1 G conditions some vestibular units do not seem to be stimulated, whereas the organ as a whole is always subjected to the earth's gravity. Gualtierotti [19] theorized that at rest the units do not discharge electric impulses when the kinocilium is either in a vertical position or detached from the sterocilia; the latter one being in line with the direction of the acceleration. In weightlessness, on the other hand, the otolith structure "floats" instead of being pushed one way or the other, depending on the state of rest or motion of the head. Hence, the decrease in electric discharge rate of isolated fibers of the vestibular nerve noted in weightlessness could very well be due to the decrease of the excitatory tonic activity associated with the zero stimulation of the otolithic organ.

At the present time, the mechanisms for the encoding of mechanical energy during the displacement of the statoconia and its transformation into an electrical process as well as their neural implications are still obscure.

3. Experiments on Human Otolith Functions

During the years 1954 through 1956, experiments were conducted at the USAF School of Aviation Medicine, Randolph Field, Texas, in which otolith functions in human test subjects were studied. The findings of two of the series of investigations, which were conducted by the author, will be discussed here, one concerning the subjective responses of the weightless man [21], and the other dealing with the oculo-agravic illusion [22]. They still have implication to human behavior and pathology in advanced spaceflight missions.

While exposed to short periods of virtual weightlessness in parabolic flight, practically all of the subjects reported sensations of rest or floating, particularly when they closed their eyes and directed their attention to the state of motion. This seemed to indicate that, in the case when the subject felt at rest, the otoliths did not register in spite of the fact that the subjects were moving at an increasing speed and high acceleration. In the second case, when the subjects reported a floating sensation, the otoliths may have discharged impulses at a rate which is characteristic of weightlessness, as Slater had hypothesized already in 1952 [23] and Gualtierotti established by his experiments. Moreover, the sensation of tumbling and of being inverted in flight was reported by several subjects. This phenomenon was also described by Russian scientists [24], and it was later called the "inversion illusion" by Graybiel and his collaborators [25, 26].

It has been known for quite some time that linear accelerations produce the so called "oculogravic illusion", which is believed to be a function of stimulation of the otolith organ. This illusion has been demonstrated in rotating devices [27, 28], in elevators [29], in aircraft [30, 31], and during linear acceleration [32, 33]. It consists of apparent motion or displacement of an object or its visual surrounding relative to the direction of acceleration. For example, if the direction of acceleration is straight ahead, an observer will perceive himself as tilted backward, and a visual target will be perceived under this condition as moving and as being displaced upward. The visual after-image of this target will move and be displayed downward, however. In weightlessness, the direction of the illusory motion is reversed. This phenomenon, which was called the "oculo-agravic illusion", was recently studied again by von Baumgarten in aircraft using a double after-image technique [34]. He concludes that both the oculogravic and the oculo-agravic illusion are partly caused by the physical conditions of the eye, in particular by eye movements, and partly by neural processes "behind the eye"; i.e. by mechanisms associated with the integration of visual and vestibular stimuli within the central nervous system.

Russian scientists also conducted experiments on vestibular functions in men [35, 36]. Results obtained on the human centrifuge showed that when $+A_z$ accelerations were applied, a nystagmic response was elicited which was thought to be associated with the increased pressure of the statoconia due to its increased weight. If the magnitude of the gravito-inertial force was decreased, the nystagmus was inhibited. That mechanical stimulation of the utricle can evoke strong primary and secondary nystagmus has been shown recently by Gernandt [37]. Earlier experiments were made by Yuganov [38], who studied the degree of excitation of the otoliths during acceleration and weightlessness by means of galvanic stimulation. The duration of the nystagmus decreased in subgravity and weightlessness.

Another effect of stimulation of the otoliths is counter-rolling of the eyes, which occurs during various forms of body tilt. Based on a series of early experiments on body positions made in the 1 G state and during accelerations on the centrifuge, Schoene [39] concluded that only in the normal body and head position have the otoliths enough input data for a relatively accurate perception of body orientation. His measurements indicated that the perceived inclination of the head or body changes proportionally to the amount of shearing force acting on the utricle. This relationship can be expressed by a simple sine function. Miller [40] also conducted experiments on tilt responses, but he used normal and labyrinthine defective subjects. By assuming a combined utriculus and sacculus response, he obtained a cosine square function as expression of lateral head tilt in normal persons. However, tilting in the sagital (pitch axis) elicited an almost straight line response as function of the field force and the utricular shear action.

Other experiments on human otolith functions were conducted by Johnson [41] and by Graybiel and Miller [42, 44] using a slow rotating room, a counter-rotating device, and an off-vertical rotating chair. The investigators assumed that the otoliths were stimulated under these conditions of rotation in a first-order fashion, although they admitted that the rotating linear acceleration vector probably affected the semicircular canals as well. As a result of this stimulation, apparent motion of a luminous target or after-image, horizontal eye movements, and motion sickness symptoms occurred, which were interpreted by the investigators as being cilio-otolithic in nature. The subject's motion sickness susceptibility, which was also studied, could be plotted as a function of the tilt of the rotating body axis, and it varied directly in accordance with Fechner's law [43]. Miller and Graybiel then hypothesized as follows:

> "If such otolith activity is the basis for the Fechnerian function found, then it would follow that the change in stressor effect that occurs with

simply a shift in off-vertical rotational position (at least within the range tested) would be dependent upon the integrated change in the amount of deformation of the macular sensory hairs, coded in logarithmic terms, in response to the constant dynamic pattern of stimulation."

Miller and Graybiel [44] also tested normal and labyrinthine defective subjects during parabolic flights. No counter-rolling response occurred during tilting in weightlessness. As the acceleration increased above zero, there was initially for the constantly held position of tilt only rather weak counter-rolling, which tended to increase at a fast increasing rate to reach the maximum at $1\,g$. This results in a logarithmic scale in accordance with the formula $\psi = \log G + b$; where ψ represents the degree of counter-rolling, G is the amount of gravity and b is a constant. A logarithmic relationship of some sort between acceleration and sensations had been predicted earlier by this and other investigators [1, 5, 45]; and this had also been confirmed by Gualtierotti's electro-physiological experiments mentioned above.

Finally, the so-called "inversion illusion" will be dicsussed, since the otoliths appear to make the primary contribution to this illusion. Subjects in aircraft during weightlessness obtained in parabolic flights and astronauts in space experienced the sensations of being rotated, and they reported the feeling that they were suspended upside down. Hence, one may wonder about the origin of the force acting on the otoconia to produce this illusion. Several authors believe that linear acceleration acting parallel to the direction of flight during the parabolic maneuver may also act on the macula, and that the direction of the force vector determines the direction of the illusory rotation leading to the "inversion sensation" [46–48].

The subjective sensation of rotation experienced under these conditions may be derived from the experiments done by Schoene [39], Jongkees [20], von Baumgarten [34] and others, who found that linear acceleration and body tilt are perceptually associated. Because the linear acceleration preceding and following weightlessness in the parabolic flight maneuver increases very gradually, the subject perceives a smooth transition from horizontal into a tilt which he interprets as a rotation. At the pull-out from the parabola essentially the reverse takes place, and the subject is quickly reoriented. Although Graybiel and Kellog [25] and Sjöberg [47] have concluded for different reasons that only the otolith response produces thc "inversion illusion", Dzendolet [48] rightly suggested that semicircular canal stimulation may also be involved.

Roman et al. [46] described the change of the directional component of the vector force in parabolic flight and gave the angle of rotation as $\psi = 7.40°$. The force vector rotates in either direction (clockwise or counter-clockwise) during the maneuver and this could induce the "backward flip" sensation reported by the subjects. Recently, Roberts [49] has stated that the illusion is caused by the rotation of the vector of the thrust, lift, and drag forces which support the aircraft during a parabolic flight maneuver. However, even if there were such a rotation of gravito-inertial forces in flight acting on the airplane, the amount of contact force acting on the weightless man should be zero. This means that in actual weightlessness the otoconia is not appressed to its surroundings; therefore, no forces are acting on the otolithic system. While there may be small residual accelerations in the aircraft, no such forces prevail within the orbiting spacecraft. Only head or body movements can there stimulate the vestibular receptor organs.

Vestibulometric studies in Voskhod 1 did not reveal any change in the threshold of sensitivity of the otoliths in the cosmonauts [38]. During the 8-day Gemini 5 and the 14-day Gemini 7 missions, ocular functions were measured on four astronauts and compared with their pre-flight and post-flight responses. While the otolith functions were depressed in space, the egocentric visual localization of the horizon was maintained [50]. Moreover, no significant change in otolithic sensitivity occurred as a result of the exposure to weightlessness.

Miller et al. [51] also conducted experiments on motion sickness susceptibility during Coriolis acceleration in parabolic flight. Susceptibility was measured by the number of head movements required to evoke severe malaise while moving on a rotating chair. In about half of their subjects, susceptibility decreased in weightlessness, in the other half, it increased as compared with the $1\,G$ threshold. The authors concluded that "any significant changes in susceptibility during weightlessness were due largely to changes in otolithic modulating influences resulting from the lifting of the gravitational load". Incidentally, Russian investigators observed a similar change of the utricular and canalicular relationship during zero G, but they warned about generalizations from this type of experiments [35, 36, 38].

4. Conclusions

Although some of the reception mechanisms of the gravireceptor in weightlessness are now well understood, certain functions and phenomena are rather controversial. Several hypotheses exist in regard to the stimulus-response relationship (logarithmic and Weber-Fechner function), the amount of deafferentiation, organ system involvement (utriculus and sacculus, contribution and role of the semicircular canals), etiology of illusory sensations, and in regard to their contribution to the motion sickness attacks observed under weightless conditions.

In weightlessness, the otolith system is deafferentiated to the extent that the pattern of certain reflexes is disturbed or extinguished and false or illusory sensations are brought about. However, this effect must not be regarded as a total deafferentiation comparable to a functional loss of the otolith organ caused by labrinthectomy or Menière's disease.

The otolith organ, deprived of the natural $1\,G$ input, responds to accelero-inertial forces, although its sensibility to such stimuli appears to be changed. There seems to be some evidence that the otolith organs exert an inhibitory influence on the semicircular canal responses in weightlessness and that the functional activity of the vestibular apparatus may thus be reduced. On the other hand, some investigators believe that the canalicular responses are less controlled by the statoconia in weightlessness and may even become more sensitive to accelero-inertial stimulations. In any case, the adaptability of the otolithic organ to the weightless condition has been established.

As to spaceflight implications, the role of the otolith organ deserves further attention. There were motion sickness attacks in all three Skylab missions, the cause of which is still not known. It is concluded at this point that the changed sensory input of the vestibular organs, in particular, the otoliths in conjunction with certain fluid shifts within the body, are intimately involved [52].

References

[1] O. Gauer and H. Haber, in: German Aviation Medicine World War II, Part I, U. S. Government Printing Office, Washington, D. C., 1950 (p. 641).

[2] S. J. Gerathewohl, Prentice-Hall Space Technology Series, Englewood Cliffs, New Jersey 1963.

[3] H. Haber, in: Physics and Medicine of the Upper Atmosphere, The University of New Mexico Press, Albuquerque, New Mexico, 1952 (p. 123).

[4] O. L. Ritter and S. J. Gerathewohl, School of Aviation Med. Rep. No. 58—154, Randolph Field, Texas, January 1959.

[5] H. Haber and S. J. Gerathewohl, J. Aviation Med. **22**, 180 (1951).

[6] A. Graybiel and J. L. Patterson, Jr, J. Appl. Physiol. **7**, 525 (1951).

[7] J. P. Henry, E. R. Ballinger, P. H. Maher and D. G. Simons, J. Aviation Med. **23**, 421 (1952).

[8] H. J. A.von Beckh, J. Aviation Med. **25**, 235 (1954).

[9] G. J. D. Schock, Aerospace Med. **32**, 336 (1961).

[10] B. G. King, Aerospace Med. **32**, 137 (1961).

[11] Ye. M. Yuganov and D. V. Afanasyev, in: Problems of Space Biology, Vol. **III**, Nauka, Moscow 1964 (p. 176).

[12] S. J. Gerathewohl and H. Stallings, J. Aviat. Med. **28**, 345 (1957).

[13] O. Gazenko and A. Kutznetzov, 2nd. World and 4th European Congr. on Aviation and Space Med. Rome, Italy, 1959.

[14] V. Fiorica, T. Semba and F. R. Steggerda, Aerospace Med. **33**, 475 (1962).

[15] R. J. von Baumgarten, Aerospace Med. **43**, 626 (1972).

[16] E. D. Adrian, J. Physiol. **101**, 384 (1943).

[17] O. Lowenstein and T. D. M. Roberts, J. Physiol. **110**, 392 (1949).

[18] T. Gualtierotti and S. J. Gerathewohl, in: NASA SP-77, Washington, D. C., 1965 (p. 221).

[19] T. Gualtierotti, NASA Contract NASW-2211, Piccin Medical Books, Milan, Italy, January 1972.

[20] L. B. W. Jongkees, in: NASA SP-152, Washington, D. C., 1968 (p. 307).

[21] S. J. Gerathewohl, Astronautica Acta **2**, 203 (1956).

[22] S. J. Gerathewohl and H. Stallings, J. Aviation Med. **29**, 504 (1958).

[23] A. E. Slater, 3rd Int. Astronaut. Congr., Stuttgart, Germany, 1952.

[24] L. A. Kitayev-Smyk, in: Problems of Space Biology, Vol. **III**, Nauka, Moscow 1964 (p. 159).

[25] A. Graybiel and R. S. Kellog, in: NASA SP-115, Washington, D. C., 1966 (p. 15).

[26] B. Clark, in: NASA SP-152, Washington, D. C., 1967 (p. 331).

[27] A. Graybiel, Arch. Ophthal. **48**, 605 (1952).

[28] A. Graybiel and B. Clark, Aerospace Med. **36**, 1173 (1965).

[29] J. I. Niven, T. C. D. Whiteside and A. Graybiel, Nav. School of Aviat. Med., Rep. No. NSAM-876, Pensacola, Fla., 1963.

[30] B. Clark, A. Graybiel and K. MacCorquodale, Am. J. Psychol. **61**, 50 (1948).

[31] J. A. Roman, B. H. Warren, J. E. Niven and A. Graybiel, Aerospace Med. **34**, 841 (1963).

[32] R. von Baumgarten, R. Thuemler, G. L. Shillinger and B. Baldrighi, 21st Int. Congr. of Aviation and Space Med. Munich, Germany, 1973.

[33] B. Clark and A. Graybiel, Am. J. Ophthal. **32**, 549 (1949).

[34] R. von Baumgarten, 5th Int. Symp. on Basic Environmental Problems of Man in Space Washington, D. C., 1973.

[35] V. I. Yazdovskiy and M. D. Yemelyanov, in: Problems of Space Biology, Vol. **III**, Nauka, Moscow 1964 (p. 79).

[36] Ye. M. Yuganov and A. J. Gorshkov, in: Problems of Space Biology, Vol. **III**, Nauka, Moscow 1964 (p. 167).

[37] B. E. GERNANDT, in: NASA SP-314, Washington, D. C., 1973 (p. 195).
[38] YE. M. YUGANOV, in: Problems of Space Biology, Vol. IV, Nauka, Moscow 1965 (p. 54).
[39] H. SCHOENE, J. Physiol. **46**, 57 (1962).
[40] E. F. MILLER II, Acta Otolaryng. **54**, 479 (1962).
[41] W. H. JOHNSON, Aerospace Med. **35**, 874 (1964).
[42] A. GRAYBIEL and E. F. MILLER II, in: NASA SP-187, Washintgon, D. C., 1968 (p. 53).
[43] E. F. MILLER II and A. GRAYBIEL, in: NASA SP-314, Washington, D. C., 1970 (p. 99).
[44] E. F. MILLER II and A. GRAYBIEL, in: NASA SP-77, Washington, D. C., 1965 (p. 121).
[45] S. J. GERATHEWOHL, J. Aviation Med. **25**, 412 (1954).
[46] J. A. ROMAN, B. H. WARREN and A. GRAYBIEL, Aerospace Med. **35**, 121 (1964).
[47] A. SJÖBERG, in: NASA SP-187, Washington, D. C., 1968 (p. 7).
[48] E. DZENDOLET, Aerospace Med. **42**, 1211 (1971).
[49] D. M. ROBERTS, Aerospace Med. **44**, 484 (1973).
[50] E. F. MILLER II, in: NASA SP-121, Washington, D. C., 1966 (p. 431).
[51] E. F. MILLER II, A. GRAYBIEL, R. S. KELLOG and R. O'DONNEL, Aerospace Med. **40**, 862 (1969).
[52] I. D. PESTOV and S. J. GERATHEWOHL, in: Foundation of Space Biology and Medicine, Vol. 2, USA/USSR Academy of Sciences (in publication).

Life Sciences and Space Research XIII — Akademie-Verlag, Berlin 1975

EFFECT OF 14 DAYS OF BED REST ON URINE METABOLITE EXCRETION AND PLASMA ENZYME LEVELS

N. PACE[a], B. W. GRUNBAUM[a], A. M. KODAMA[a], D. F. RAHLMANN[a] and B. D. NEWSOM[b]

[a] White Mountain Research Station, University of California, Berkeley, USA
[b] NASA Ames Research Center, Moffett Field, California, USA

After one week of ambulatory base-line measurement, a group of eight men 19—26 years of age remained continuously recumbent for 14 days. Studies were continued for one week following the prolonged recumbency. Urine excretion rates for a number of constituents were determined 2 days before bed rest, on day 14 of bed rest, and day 6 after bed rest. Blood plasma samples were also obtained at these times, and analyzed for several enzymes. On day 14 of bed rest significant increases were observed in urine excretion of total osmotically active substances, magnesium, calcium, phosphate, creatinine, hydroxyproline, and 17-OH corticosteroids. A decrease occurred in urinary glucose excretion. Plasma levels of alkaline phosphatase and LDH-3 were depressed, while plasma GPT was elevated. Many of these changes persisted on day 6 after bed rest, and are interpreted as concomitants of the disuse atrophy of the musculoskeletal system that characterizes prolonged bed rest and weightlessness.

1. Introduction

In the course of a larger study conducted at the Human Research Facility of the NASA Ames Research Center, the effects of 14 days of bed rest were examined on eight male subjects. The present paper deals with the results of measurement of the urinary excretion rates of a number of metabolites, and of the level of several plasma enzymes in the course of the study. Results from other aspects of the study are published elsewhere. The parameters reported here reflect primarily the metabolic status of the musculoskeletal system before and after a period of prolonged recumbency, and may provide useful indication of the changes to be expected from exposure to the weightlessness of space flight.

2. Procedures and Methods

The eight subjects examined were healthy student volunteers in the age range 19 to 26 years. During the 1st week of the study the subjects remained ambulatory to provide base-line data before the start of the bed-rest period. The subjects then remained continuously recumbent for 14 days. Eating up on one elbow, excreting, and bathing were all accomplished horizontally. A nutritionally balanced diet of up to 2800 kcal daily was provided throughout the study. In the 4th week of the study the subjects were once again ambulatory to provide measurements during the recovery period.

Four of the subjects (FIR, SMI, HAR and HEI) wore a special reverse-gradient garment for 4 hours daily during the bed-rest phase of the study. The garment was designed to force blood into the lower limbs by a combination of positive-pressure breathing and an external pressure gradient diminishing toward the feet, supplied by elastic fabric to the lower torso and limbs. However, for all of the parameters reported here, and for most in the study as a whole, a statistically significant difference between the response of these four subjects to bed rest and that of the other four could not be detected by analysis of variance. Hence, the results from all eight subjects were pooled.

A 24-hour urine collection was made starting on day 2 before the onset of bed rest, on day 14 of bed rest, and on day 6 after the end of bed rest. A post-absorptive venous blood sample was obtained in the morning of those days.

Urine parameters measured were: volume, specific gravity, total osmotic activity by freezing-point depression, pH, chloride, sodium, potassium, magnesium, calcium, phosphate, total nitrogen, ammonia, urea, creatinine, creatine, hydroxyproline, glucose, citrate, 17-hydroxycorticosteroids, epinephrine, norepinephrine, and cyclic adenosine monophosphate (AMP). Epinephrine and norepinephrine were determined spectrophotofluorometrically using a kit supplied by Rio-Rad Laboratories, Richmond, California, based on the method of Sandhu and Freed [1]. Cyclic-AMP was measured by competitive protein binding, using a kit supplied by the Amersham-Searle Corporation, Arlington Heights, Illinois, based on the method of Gilman [2]. All the other urine constituents were analyzed by the methods of Grunbaum and Pace [3].

The heparinized blood samples were centrifuged to separate the plasma for analysis. Alkaline phosphatase (ALP) concentration was determined using a kit supplied by Hycel, Inc., Houston, Texas, based on the method of Bodansky and Schwartz [4]. Glutamate-oxalacetate transaminase (GOT) and glutamate-pyruvate transaminase (GPT) levels were measured using a kit supplied by the Dade Company, Miami, Florida, based on the method of Reitman and Frankel [5]. Total lactate dehydrogenase (LDH) concentration was determined using a kit supplied by the Dade Company, based on the method of Cabaud and Wroblewski [6]. The individual lactate dehydrogenase isoenzymes were separated electrophoretically and measured by the method of Grunbaum et al. [7].

The data obtained were examined by the two-way analysis of variance for paired comparisons [8]. A value of $P < 0.05$ was taken to indicate a statistically significant difference between the two group means being compared.

3. Results

As may be seen in Table 1*, a statistically significant increase occurred in urine specific gravity and the amount of osmotically active material excreted during 24 hours on day 14 of bed rest as compared with the values measured 2 days before bed rest began. The 24-hour urine volume and pH were not changed significantly on day 14 of bed rest as compared with the pre-bed-rest values. All four of these parameters were essentially the same on day 6 after bed rest as on day 2 before bed rest.

---

* Means only are given in the Tables. Full details of individual values are available from the authors on request.

Table 1
Urine 24-hr Values for Volume, Solids and pH (means of all subjects)

Time rel. to Bed Rest	Volume (liters)	Specific Gravity	Osmotic Activity (osmols)	pH
2 days before	2.18	1.013	1.08	6.5
Day 14	1.78	1.018*	1.18*	6.5
6 days after	1.89	1.015*	1.02	6.4

* Value significantly different ($P < 0.05$) from the corresponding mean obtained 2 days before bed rest.

Table 2
Urine 24-hr Values for Electrolyte Excretion (means of all subjects)

Time rel. to Bed Rest	Chloride (mmols)	Sodium (mmols)	Potassium (mmols)	Na/K Ratio	Magnesium (mmols)	Calcium (mmols)	Phosphate (mmols)
2 days before	206	198	81	2.50	4.46	3.74	34.4
Day 14	226	198	74	2.80	5.58*	5.82*	41.5*
6 days after	181	172	75	2.56	4.25	4.81*	43.8*

* Value significantly different ($P < 0.05$) from the corresponding mean obtained 2 days before bed rest.

Table 3
Urine 24-hr Mean Values for Nitrogen Excretion (means of all subjects)

Time rel. to Bed Rest	Total N (g)	Ammonia (mmols)	Urea (mmols)	Creatinine (mmols)	Creatine (mmols)	Hydroxy-proline (mmols)
2 days before	14.8	34.4	559	15.3	1.70	0.503
Day 14	16.3	38.2	611	16.3*	2.57	0.675*
6 days after	14.7	41.5*	608	15.6	1.97*	0.520

* Value significantly different ($P < 0.05$) from the corresponding mean obtained 2 days before bed rest.

The excretion rates for the major urinary electrolytes are shown in Table 2*. Chloride, sodium and potassium excretion rates were not demonstrably different on day 14 of bed rest as compared with day 2 before bed rest. In contrast, magnesium, calcium and phosphate excretion rates were strikingly elevated. Magnesium excretion rate had returned to the base-line value by day 6 after bed rest, but both calcium and phosphate excretion rates were still significantly elevated.

Table 3* shows the measurements obtained for the principal nitrogenous con-

* Means only are given in the Tables. Full details of individual values are available from the authors on request.

stituents of urine. There was a marked tendency for total nitrogen excretion rate to be greater on day 14 of bed rest, as compared with day 2 before bed rest, in that seven of the eight subjects showed an increase. However, this trend did not achieve statistical significance. On the other hand, it may be seen that creatinine excretion rate was elevated slightly but significantly on day 14 of bed rest, and that hydroxyproline excretion was substantially elevated. Both of these parameters had returned to base levels by day 6 after bed rest; however, at this time ammonia and creatine excretion rates appeared to be increased significantly.

Values for several other urine constituents are given in Table 4*. Of these, glucose excretion rate was significantly depressed both on day 14 of bed rest and on day 6 after bed rest. The excretion rate of 17-hydroxycorticosteroids was

Table 4

Urine 24-hr Mean Values for Excretion of Various Metabolites (means of all subjects)

Time rel. to Bed Rest	Glucose (mmols)	Citrate (mmols)	17-OHCS (μmols)	Epinephrine (nmols)	Norepinephrine (nmols)	Cyclic-AMP (nmols)
2 days before	0.89	1.18	2.34	37.4	215	1.18
Day 14	0.59*	1.61	5.13*	21.9	318	0.88
6 days after	0.45*	1.69	1.57	16.0	282	1.02

* Value significantly different ($P < 0.05$) from the corresponding mean obtained 2 days before bed rest.

Table 5

Plasma Enzyme Concentration (means of all subjects)

Time rel. to Bed Rest	ALP (units/ml)	GOT (units/ml)	GPT (units/ml)	Total LDH (units/ml)
2 days before	31.7	40.3	24.5	303
Day 14	17.0*	42.3	31.2*	291
6 days after	21.3*	40.4	26.9	290*

* Value significantly different ($P < 0.05$) from the corresponding mean obtained 2 days before bed rest.

considerably increased on day 14 of bed rest, but had returned to base-line level by day 6 after bed rest. Excretion rates of citrate, epinephrine, norepinephrine, and cyclic-AMP were not discernibly different on day 14 of bed rest or day 6 after bed rest.

Blood plasma concentrations of four enzyme systems are shown in Table 5*. Alkaline phosphatase and lactate dehydrogenase levels were both depressed significantly, and glutamate-pyruvate transaminase level was elevated signif-

* Means only are given in the Tables. Full details of individual values are available from the authors on request.

icantly on day 14 of bed rest. Glutamate-oxalacetate transaminase level was unchanged. Glutamate-pyruvate transaminase had returned to base-line values on day 6 after bed rest, but both alkaline phosphatase and lactate dehydrogenase remained significantly depressed.

The five isoenzymes of lactate dehydrogenase were separated electrophoretically and measured separately, with the result seen in Table 6*. The slight fall in

Table 6
Plasma Lactate-Dehydrogenase Isoenzyme Concentrations (means of all subjects)

Time rel. to Bed Rest	LDH-1 (units/ml)	LDH-2 (units/ml)	LDH-3 (units/ml)	LDH-4 (units/ml)	LDH-5 (units/ml)
2 days before	102	95	58	22	26
Day 14	104	84	48*	21	33
6 days after	93	84*	58	27*	29

* Value significantly different ($P < 0.05$) from the corresponding mean obtained 2 days before bed rest.

total lactate dehydrogenase concentration noted on day 14 of bed rest seemed to be accounted for by a fall in LDH-3 concentration, the other isoenzyme levels remaining unchanged. On day 6 after recovery small differences from the pre-bed-rest values were observed in the LDH-2 and LDH-4 levels. It is worthy of note, however, that neither the LDH-1 nor the LDH-5 isoenzymes showed a significant change after 14 days of bed rest.

4. Discussion

The total pattern of change in the parameters measured is summarized in Table 7. It is clear that after 2 weeks of bed rest, urine volume production and urine pH are the same as in the normal ambulatory state. However, there is indication in the elevated specific gravity and excretion rate of osmotically active substances of a greater loss of solutes from the body at this time.

When the individual urinary electrolyte excretion rates are examined, it may be seen that magnesium, calcium, and phosphate excretion were all greater on day 14 of bed rest than before bed rest. Magnesium excretion rate returned to base-line levels within 6 days after bed rest, but calcium and phosphate excretion rates remained persistently higher. These results for calcium and phosphate conform to those seen in the classic bed-rest study of Deitrick et al. [9]. They also resemble closely the findings of Whedon and Lutwak [10] on the Skylab 1/2 astronauts, which included measurement of urinary magnesium excretion rate. There seems little doubt, therefore, that both bed rest and weightlessness result in an increased urinary electrolyte excretion rate.

It is worthy of note that the plasma alkaline phosphatase levels were significantly reduced on day 14 of bed rest, and that the lower levels persisted for at least

* Means only are given in the Tables. Full details of individual values are available from the authors on request.

Table 7

Summary of Statistically Significant ($P < 0.05$) Changes in Urine Excretion Rate and Plasma Level of the Parameters Studied

Direction of the arrows indicates an increase (↑) or a decrease (↓), while zero indicates no change from the values measured 2 days before bed rest.

Parameter	Day 14 of bed rest	Day 6 after bed rest	Parameter	Day 14 of bed rest	Day 6 after bed rest
Urine			*Urine*		
Volume	0	0	Glucose	↓	↓
Specific gravity	↑	↑	Citrate	0	0
Osmotic activity	↑	0	17-OHCS	↑	0
pH	0	0	Epinephrine	0	0
Chloride	0	0	Norepinephrine	0	0
Sodium	0	0	Cyclic-AMP	0	0
Potassium	0	0			
Na/K ratio	0	0	*Plasma*		
Magnesium	↑	0	Alk-phosphatase	↓	↓
Calcium	↑	↑	GOT	0	0
Phosphate	↑	↑	GPT	↑	0
Total nitrogen	0	0	Total LDH	↓	↓
Ammonia	0	↑	LDH-1	0	0
Urea	0	0	LDH-2	0	↓
Creatinine	↑	0	LDH-3	↓	0
Creatine	0	↑	LDH-4	0	↑
Hydroxyproline	↑	0	LDH-5	0	0

6 days after bed rest. Furthermore, the urinary hydroxyproline excretion rate was markedly elevated on day 14 of bed rest. It is known that plasma alkaline phosphatase concentration correlates with activity of the osteoblasts in forming new bone, so that a decrease in alkaline phosphatase levels implies a reduction of skeletal osteoblast activity. On the other hand, urinary hydroxyproline excretion rate correlates with catabolic loss of organic bone matrix. Thus, the implication is clear that during bed rest there is a net loss of skeletal mass of the body, partly through a reduced rate of formation of new bone and partly through an increased catabolic loss of bone substance. The transient is apparently a slow one, requiring at least several weeks to reach completion.

The other area of interest is the indicators of the status of the skeletal musculature. On day 14 of the present bed-rest study there was the suggestion of an elevated total nitrogen excretion by way of the urine, a finding reported both by Deitrick et al. [9] and by Whedon and Lutwak [10]. There was also indication of a small increase in urinary creatinine, creatine, and ammonia excretion rates. It may be of significance in this connection that total plasma lactate dehydrogenase concentration was depressed somewhat at the end of the bed-rest period. However, the change in total plasma LDH level seemed to be a consequence of reductions in LDH-2 and LDH-3 isoenzyme concentrations, while LDH-1 and LDH-5 levels were unchanged. LDH-1 arises primarily from heart muscle, and LDH-5 arises primarily from skeletal muscle. Thus, one would have expected to see some change in these isoenzyme levels as a correlate of the reduced muscular activity during bed rest. There is no ready explanation for the increase in glutamate-pyruvate transaminase levels observed on day 14 of bed rest, inasmuch as this plasma enzyme is thought to arise primarily in the cells of the liver. In any case, the results from the present study do not contradict the conclusion by other investigators that loss of skeletal muscle mass occurs during bed rest, and tend to support it.

Brief comment may be made concerning the significant decrease in urinary glucose excretion rate which was observed at the end of the bed-rest period. This may be correlated with the increase in urinary 17-hydroxycorticosteroid excretion rate observed concomitantly, and an expected increase in hepatic glycogenesis. A finding of increased urinary cortisol excretion rates in the Skylab 1/2 astronauts was reported by Leach [11], and there is little doubt of the importance of further study of the endocrine interrelationships in both bed rest and weightlessness.

In general, the present study demonstrates the utlity of the metabolite pattern approach to the study of the disuse atrophy of the musculoskeletal system that accompanies prolonged periods of bed rest and weightlessness. It is through such an approach that the physiological adjustments involved can eventually be understood completely, and perhaps be prevented or reversed at will.

Acknowledgments

The work performed during this study and travel to attend the COSPAR meeting where it was presented were supported by NASA Grant NGR 05-003-470. Thanks are due the many other members of the staffs of these two laboratories, and the subjects of the study, for the considerable help they provided.

References

[1] R. S. Sandhu and R. M. Freed, Standard Methods of Clinical Chemistry, vol. 7, Academic Press, New York 1972 (p. 231).

[2] A. G. Gilman, Proc. Nat. Acad. Sci. USA **67**, 305 (1970).

[3] B. W. Grunbaum and N. Pace, Microchem. J. **9**, 184 (1965); **15**, 666 (1970); **15**, 673 (1970); **16**, 443 (1971); **18**, 146 (1973).

[4] O. Bodansky and M. K. Schwartz, Methods in Medical Research, vol. 9, Year Book Publishers, Chicago 1961 (p. 79).

[5] S. Reitman and S. Frankel, Am. J. Clin. Path. **28**, 56 (1957).

[6] P. G. Cabaud and F. Wroblewski, Am. J. Clin. Path. **30**, 234 (1958).

[7] B. W. Grunbaum, J. Zec and E. L. Durrum, Microchem. J. **7**, 41 (1963).

[8] R. R. Sokal and F. J. Rohlf, Biometry, W. H. Freeman and Co., San Francisco 1969 [p. 328).

[9] J. E. Deitrick, G. D. Whedon and E. Shorr, Am. J. Med. **4**, 3 (1948).

[10] G. D. Whedon and L. Lutwak, paper presented at 5th Int. Symp. on Basic Environmental Problems of Man in Space, Washington, D. C., November 1973.

[11] C. S. Leach, paper presented at 5th Int. Symp. on Basic Environmental Problems of Man in Space, Washington, D. C., November 1973.

Life Sciences and Space Research XIII — Akademie-Verlag, Berlin 1975

INFLUENCE OF SIMULATED WEIGHTLESSNESS ON THE RATE OF ANOMALIES OF THE FLOUR BEETLE *Tribolium confusum*

W. Briegleb, J. Neubert, A. Schatz and F. Sinapius
Technical Assistance, F. Schuster

DFVLR, Institut für Flugmedizin, Bonn-Bad Godesberg, FRG

Experiments with *Tribolium confusum* showed that the morphological characteristics of the beetles are modified by simulated weightlessness (fast running clinostat). Because of possible side effects due to differences in fertility of inbred lines, the first experiments were made with a genetically heterogeneous stock. Thereafter experiments were continued with inbred beetles. For both stocks a rise of mainly wing anomalies resulted from rotation of whole cultures of beetles within horizontal tubes. The extent to which these anomalies are teratogenetic or genetic has not yet been analysed in detail.

1. Introduction

Whether weightlessness has mutogenic or teretogenic effects has not yet been clarified. Certain results from the Gemini flights and Biosatellite 2 led to different interpretations [1]. By using a fast clinostat [2] we could demonstrate that simulation of weightlessness does not influence ontogenesis of the flour beetle *Tribolium casteneum* [3] at least until hatching from the egg.

2. Method

In a viscous medium which rotates rather fast about a horizontal axis, heavy particles near the centre of rotation are forced to behave much as in free fall [4—6]. Side effects of rotation such as tumbling of heavier particles and centrifugation are masked by Brownian movement [7]. We assumed that there may be a higher acceleration threshold for perception by hypothetic reactions sensitive to gravity which, unlike statocysts, are not specialized for gravity perception. Because of the dimension of the beetle (4 mm length) the area of experimentation must be larger than in the case of small single cells.

Two different genetic stocks of the beetle *Tribolium confusum* were used: a genetically heterogeneous stock and an inbred line derived from it. The inbred line (from several successive brother-sister matings) was selected to eliminate all visible morphological defects.

The beetles of the main experiment were cultured within small plexiglass tubes (28 cm long, 4 mm internal diameter; Fig. 1), rotating longitudinally at 90 revolutions min^{-1}. Controls were set up in non-rotating tubes (tube controls) and in normal culture vessels (normal controls). Once a generation, the cultures were harvested and the number of animals within the tubes reduced to 8 (whether larvae, pupae or adults). After the sixth or eighth harvests the experiments were stopped and new experiments started with wild-type animals from stock cultures. One rotor and one control tube were started with st (split elytra) animals. The anomaly st is a phenotypic trait whose genetic basis is not yet understood. The

Fig. 1. Two of the four rotors maintaining small populations of the beetle *Tribolium confusum*. The inner tubes of the rotors are 28 cm long and 4 mm inner diameter. In each generation each stock is reduced to 8 animals and is provided with new nourishment. Temperature is kept between 25 and 30 °C with relative humidity 40–50%. Rotation speed is 90 rev min^{-1}.

st animals used to start rotating tubes were animals from rotated cultures and those used to start control cultures were animals from control cultures. The harvested animals from each generation were cultured in different ways as groups or starting with pairs. About 80 crosses have been made to analyse the genetic behaviour of single anomalies (for further details of the method see [8]), but detailed results are not yet available.

3. Results

During our work we observed a number of different anomalies which are listed in Table 1. The abbreviations used describe the characters in a preliminary way, and some of them may be identical with known genes in the literature. For comparability with the findings of Buckhold et al. [9] on the Biosatellite 2 experiment we used the st anomaly in the variations and combinations listed in Table 1 to demonstrate changed levels of malformations within our cultures. The frequencies of the other anomalies observed essentially parallel the frequencies of st.

Table 2 shows that the frequency of abnormalities is greatly increased in the rotated cultures and shows also the probability levels (based in part on χ^2 and

Table 1

Phenotypic Anomalies found

w	wildtype	ia	incomplete, fringed alae
st	split elytra	ba	blistered alae
hst	elytra split on one half side	py	pygmy
sth	elytra split behind	g	giant
stm	elytra split in the middle	ua	upturned abdomen
tet	thickened elytral tips	tu	tumours of the epidermis
tst	twisted split elytra	iex	incomplete extremities
be	bent elytra	wf	extremities without function
te	twisted elytra	fp	faults in pigmentation
le	lengthened elytra	es	extruded genitals
se	shortened elytra	da	dark contraction near the after part of the abdomen
ie	incomplete, fringed and/or depigmented elytra	aa	agglutination near the after part of the abdomen
bse	blistered elytra	ap	attached pupal skin (also on the legs)
ame	asymmetrical malformation of elytra	fe	extremities and/or wings fettered by the pupal skin
cd	curved decay with dark margin within the epidermis	im	incomplete metamorphosis
pab	alae protruded between elytra	rv	reduced viability
pao	alae protruded outside elytra		

Table 2

Statistical Significance for Differences in the Number of Wing Anomalies between Experiments and Controls

	No. of Adults		No. of Anomalies		% Anomalies		P of
	Controls	Expts.	Controls	Expts.	Controls	Expts.	Difference [11]
Harvests I—VI w* ancestors	377	565	16	81	4.25	14.33	>99.73
Harvests VII—VIII w ancestors	214	121	10	26	4.67	21.45	>99.73
Harvest X inbred w ancestors	111	127	5	23	4.5	18.1	>99.73
Harvests VII—VIII st ancestors	57	118	1	32	1.76	27.05	>99.73
Progeny from Harvest V	4403	4685	160	435	3.64	9.28	≫99.73
Progeny from Harvest VI	640	400	29	54	4.53	13.5	>99.73
Character se Progeny from Harvest V	7659	6601	5	34	0.001	0.01	≫99.73

* w = wild-type

F tests) for the differences of frequencies. When experiments were started with beetles of st phenotype, the frequency fell from the initial 100%, but fell more in the control than the rotated tube. But when beetles from these experiments are harvested and cultured in normal conditions, the resulting progeny have very similar frequencies of abnormalities. This is unlike the results with wild-type beetles, in which the progeny from rotated cultures show more anomalies (10–20%) than from unrotated cultures (about 5%).

A different result was obtained with the inbred stock. Again there was a significant difference in the proportion of anomalies in the harvested beetles (Table 2), but the number of anomalies in progeny of the rotated cultures and the tube controls was more like that of the wild-type. The descendants of both the rotated normal and inbred stocks show a clear decrease of viability. The frequency of anomalies of some other types is also elevated in the rotated experiments. A significant rise in "short elytra" (se) was found in the progeny of rotated beetles (Table 2). A dwarf form (similar to "tiny" in the literature) and leg reduction may also occur.

4. Discussion

The main question is whether the observed frequency changes of anomalies within our cultures really result from the simulation of weightlessness. Another question is what is their physiological and genetic origin. One example of a physiological effect is the so-called quinone effect, which involves a chemical product of the beetle itself and Sokoloff [10] speaks of legs anomalies in this connection. A comprehensive interpretation of the genetical data is not yet possible, but it may be that the phenotypic character st is of polygenetic origin and incomplete penetrance. We are continuing to investigate this problem. It has to be remembered that selection of genes for anomalies may have occurred as a result of rotation, both in the heterogeneous and inbred strains.

References

[1] J. F. Saunders et al., in: The Experiment of Biosatellite II, NASA SP-204, Washington 1971 (p. 1).
[2] W. Briegleb, Naturwissenschaften **54**, 167 (1967).
[3] W. Briegleb, in: Proc. XVIIIth Int. Astronaut. Congr., Belgrade 1967, Vol. 4: Life in Spacecraft, Pergamon Press, Oxford 1968 (p. 87).
[4] W. Briegleb et al., in: Proc. XXth Int. Astronaut. Congr. Mar del Plata 1969, Part 4: Bioastronautics, Pergamon Press, Poland 1972 (p. 859).
[5] A. Schatz et al., in: Aktuelle Forschungsarbeiten aus dem Institut für Flugmedizin, DLR-FB 73-15, Porz-Wahn 1973 (p. 169).
[6] A. Schatz et al., in: Preprints of XXIst Int. Congr. of Aviation and Space Medicine, München 1973 (p. 111).
[7] W. Briegleb et al., in: Preprints of XXIst Int. Congr. of Aviation and Space Medicine, München 1973 (p. 113).
[8] W. Briegleb et al., in: Aktuelle Forschungsarbeiten aus dem Institut für Flugmedizin, DLR-FB 73-15, Porz-Wahn 1973 (p. 7).
[9] B. Buckhold et al., in: The Experiment of Biosatellite II, NASA SP-204, Washington 1971 (p. 79).
[10] A. Sokoloff, The Genetics of Tribolium and Related Species, Academic Press, New York, London 1966.
[11] S. Koller, in: Biochemisches Taschenbuch, II. Teil, Springer Verlag, Berlin, Göttingen, Heidelberg, New York 1964 (p. 959).

Life Sciences and Space Research XIII — Akademie-Verlag, Berlin 1975

THE DEVELOPMENT OF SEEDLING SHOOTS UNDER SPACE FLIGHT CONDITIONS

A. J. MERKYS, A. L. MASHINSKY, R. S. LAURINAVICHIUS, G. S. NECHITAILO, A. V. YAROSHIUS and E. A. IZUPAK

Plant Physiology Laboratory, Institute of Botany, Academy of Sciences of the Lithuanian S. S. R., USSR

The assumption that gravity is the major factor in the process of formation of plant polar axis was used as a working basis for the experiment. It was hypothesized by Merkys in 1973 that the effect of gravity related to axial polarity is similar to the process which determines the lateral polarity of shoots under the influence of gravity.

There are two possibilities: (i) the development and morphogenesis of shoots takes place directly under the influence of gravity, or (ii) this process, at least during the first growth phases, is determined in the course of the germ development in the seed. In accordance with that assumption, the experiment was carried out in 1973. A special system was used for germinating and cultivating "Pioneer" and "Grybovsky rannyj" peas. The duration of the experiment under flight conditions was 48 hours in darkness, at 20 °C. The experimental conditions were the following: 1, the experiment in flight; 2, imitation of flight conditions using the horizontal clinostat; 3, vertical clinostat; 4, control (vertical plants).

When the system was brought down to earth, the material was fixed and subjected to morphological and biochemical analysis. On the basis of the analysis, the following conclusion was drawn: during the first growth phases, the morphogenesis of shoots and roots apparently does not change under flight conditions. This conclusion was confirmed by planting those seedlings under earth conditions; normal plants were obtained whose growth and development were similar to the control seedlings. The problem of the influence of changed gravity, or the lack of it, on the growth and development of plants is discussed. The possible role of gravity on the formation of the polarity axes is also discussed from the point of view of generative development and the determination of some peculiarities of morphogenesis.

1. Introduction

Earlier investigations have been carried out on growth, orientation and biochemical response of monocotyledon seedlings under conditions of space flight [1—4]. These tests (under residual gravity of only 10^{-5} g) showed no significant growth differences, except for loss of space orientation.

We hypothesized that gravity may have a decisive effect in the formation of polarity axes of the shoot, as is the case for the formation of lateral polarity of the shoot, under the geotropic reaction [5]. The organization of the experiment under conditions of weightlessness was based on the alternative hypothesis, that the formation of growth polarity and order of morphogenesis in seedlings of higher plants are determined directly by gravity during germ formation on the maternal plant. This influences first phases of both the growth and the morphogenesis of the seedling. Hence, it was expected that conditions of weightlessness would not prevent germination and seedling growth. The experiments were aimed at determining the first phases of germination and morphogenesis of pea seedlings under space flight conditions.

2. Materials and Methods

The investigation was carried out in 1973. Peas of the variety "Pioneer" and "Grybovsky rannyj" were used, under the following scheme:

1. Experiment under flight conditions
2. Imitation of flight conditions (horizontal clinostat 1 rev min^{-1})
3. Vertical clinostat (1 rev min^{-1}).
4. Control (vertical plants).

The preliminary preparations included: soaking of sterile seeds for 4 hours in distilled water at $20 \pm 2\,°C$; permitting swelling and the beginning of germination to occur for 21 hours 30 minutes at the same temperature; planting on a sterile gauze matrix. After that the seeds on the matrix were placed in the sterile container (inner volume 550 cm^3). The container, in its turn, was placed in the thermostat. The thermostat was maintained at $\sim 5°C$, which retards germination before the beginning of the experiment. The experiment itself was carried out at 23 °C and lasted 23 hours 50 minutes. Subsequently, the second experiment was carried out, using the same scheme, and differed only in that seedlings were in space flight for a longer period, i.e. 48 hours. One hour and 15 minutes after the experiment the seedlings were fixed with the Carnoy fixative. The fixed material was used later on for the morphological and biochemical analysis.

3. Results

The seed germination in all four versions of the test was 100%. The movement of the seedlings in all versions of both pea varieties did not differ significantly, except that in the test lasting 24 hours the roots were smaller. Fig. 1 shows apices of the variety "Pioneer", which grew in space flight for 48 hours, as well as those of the controls. The figure shows that a number of the first internodes of the shoot are already determined in the embryo of the ungerminated seed. In space flight, as well as during imitation of weightlessness on the clinostat, shoots grew in such a way that their internodes, especially the first and the second, began to elongate noticeably. The difference between the control and flight experiments in the root diameter and in the number of cell rows, is, however, statistically not significant.

The length of cells of the extension zone was greater under flight conditions (Table 1). The length of roots under these conditions was also greater (Table 2). In the other zones of the root apex cells did not differ significantly from the control. The root cap cells were more extended in seedlings in space flight and in seedlings which grew on the horizontal clinostat.

In all versions of the test amyloplasts and statolytes had formed in the root caps. The formation of statolytes in wheat seedlings during space flight has been noted before [1].

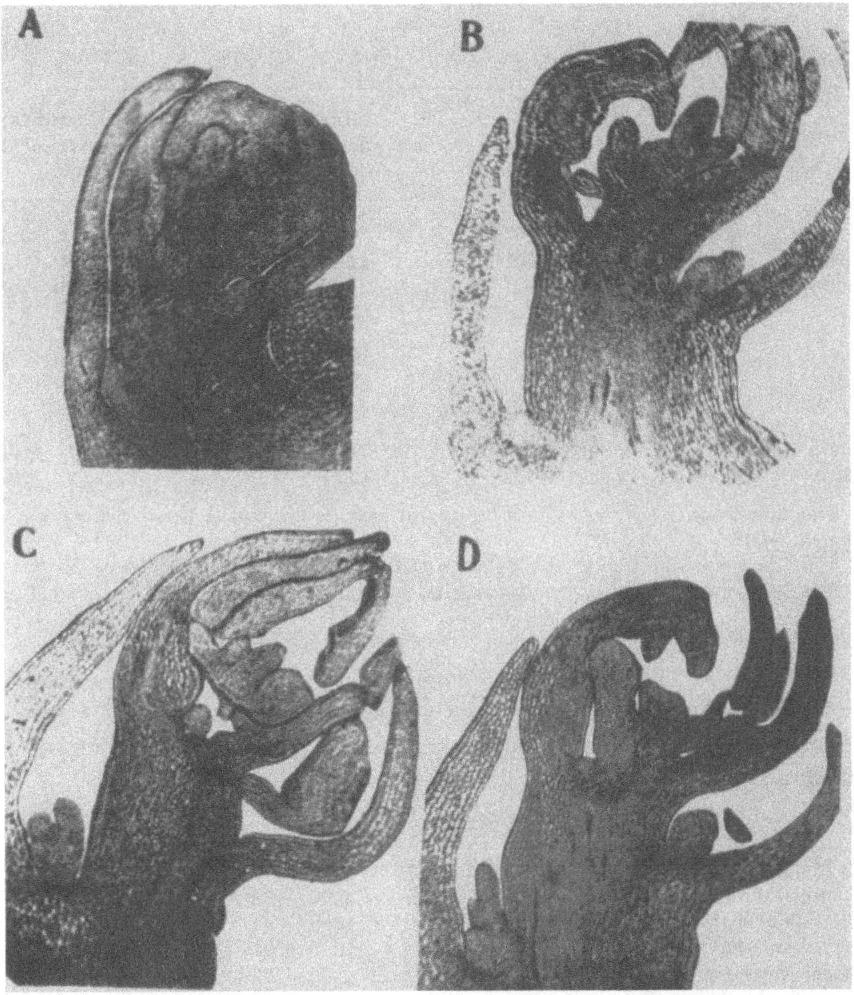

Fig. 1. Apices of pea seedlings (shoots): A, apex of the embryo shoot before germination; B, apex of the shoot grown under flight conditions; C, apex of the shoot grown on the horizontal clinostat; D, apex of the shoot grown on the vertical clinostat. Magnification 18 ×.

The rate of protein hydrolysis in cotyledons during germination after one day was estimated by the quantity of residual protein, determined by the Khjeldahl method. A tendency toward an increase in residual protein in cotyledons under flight conditions and in seedlings which grew on the horizontal clinostat was observed (Table 3), but the differences were within the limits of assay for protein.

The amount of hydrolysis of various protein fractions also showed no differences in the several experiments, using ion exchange chromatography of separate amino acids.

Table 1
Characteristics of the Root Cell Growth of the Pea "Pioneer" ($\bar{x} \pm s_{\bar{x}}$ in μm)[a]

	Flight conditions	Imitation of flight conditions (horizontal clinostat)	Control (vertical plants)
	Extension zone (2 mm from root end)		
Length	28.30 ± 0.57[c]	27.10 ± 0.58[b]	24.60 ± 0.53
Width	24.40 ± 0.59	26.00 ± 0.59	25.90 ± 0.48
	Meristem zone		
Length	15.31 ± 0.42	14.70 ± 0.34	15.30 ± 0.29
Width	16.30 ± 0.43	17.30 ± 0.34	17.10 ± 0.30
	Root cap		
Length	27.90 ± 0.61[b]	29.51 ± 0.54[c]	25.10 ± 0.53
Width	12.63 ± 0.53	12.11 ± 0.55	12.72 ± 0.47

[a] The data on only one variety are presented, because differences between varieties were not significant.
[b] The difference from the control is statistically significant at $P = 0.99$.
[c] The difference from the control is statistically significant at $P = 0.999$.

Table 2
Growth of Pea Roots during Space Flight and under Control Conditions ($\bar{x} \pm s_{\bar{x}}$ in mm)

Experiment	Root length	
	"Pioneer"	"Grybovsky rannyj"
Flight conditions	5.00 ± 0.25[a]	5.20 ± 0.25[b]
Imitation of flight conditions (horizontal clinostat)	4.80 ± 0.27	4.51 ± 0.23
Vertical clinostat	4.70 ± 0.37	4.50 ± 0.28
Control (vertical plants)	4.00 ± 0.31	4.60 ± 0.29

[a] The difference from the control is statistically significant at $P = 0.95$
[b] The difference from the control is statistically significant at $P = 0.99$

Table 3
The Influence of Space Flight Conditions on the Quantity of Protein in Pea Cotyledons of the variety "Pioneer"
(mg g^{-1} of dry weight)

Germination conditions	Protein
Flight conditions	221.2
Imitation of flight conditions (horizontal clinostat)	233.9
Vertical clinostat	231.7
Control (vertical plants)	212.6

In summary, germination, growth and the morphogenesis of the shoot under conditions of space flight takes place without significant change from normal growth. As far as the experiments go (the first phases of the germination took place under ground conditions) these data support the opinion that the first phases of the growth are already determined in the embryo and do not need the action of gravity for successful morphogenesis of the seedling and root.

References

[1] S. W. Gray and B. F. Edwards, in: The Experiments of Biosatellite II, NASA, Washington, D. C. 1971 (p. 123).
[2] C. J. Lyon, in: The Experiments of Biosatellite II, NASA, Washington, D. C. 1971 (p. 167).
[3] H. M. Conrad, in: The Experiments of Biosatellite II, NASA, Washington, D. C. 1971 (p. 189).
[4] S. P. Johnson, in: The Experiments of Biosatellite II, NASA, Washington, D. C. 1971 (p. 213).
[5] A. Merkys, Geotropic Reaction of Plants, Vilnius 1973.

Life Sciences and Space Research XIII — Akademie-Verlag, Berlin 1975

CORRESPONDENT GROUP FOR NOMENCLATURE AND TERMINOLOGY IN GRAVITATIONAL BIOLOGY

During the 1973 Business Meeting of Panel 5.A on Gravitational Biology, as part of the 16th Plenary COSPAR Meeting in Konstanz, it was decided to develop an official nomenclature and terminology in gravitational biology. The mechanism suggested for implementation was to identify the widest possible correspondent group of interested individuals in the field and arrive at a consensus position for eventual publication. Professor Arthur H. Smith, U. S. A., was appointed Chairman of the Correspondent Group to coordinate its activities.

In the past year, Professor Smith in collaboration with Professor Nello Pace, Chairman of Panel 5.A, generated a list of 80 names of individuals potentially interested. Of these, 70 persons from 13 nations have indicated a willingness for active participation in response to an initial mailing by Professor Smith. On the basis of the responses, a preliminary outline delineating the field of gravitational biology was developed, and was presented by Professor Smith at the 1974 Business Meeting of Panel 5.A during the 17th Plenary COSPAR Meeting in São Paulo. Agreement was reached among the members present that this outline, shown below, is a viable starting point for the definition of nomenclature and terminology.

I. *Geogravics*

- A. Gravity Transduction
  - 1. Animal
    - a. Geotropism and Posture
      - i. Vestibular Function
      - ii. Proprioceptor Function
    - b. Embryological Development
  - 2. Plant
    - a. Statolith Function
    - b. Geotropism
  - 3. Similitude (Scale Effects)
- B. Field Orientation
  - 1. Recumbency
    - a. Tilt Table
    - b. Bed Rest
  - 2. Clinostat

C. Loading
   1. Increased (Weighting)
   2. Decreased (Counterbalancing)

D. Activity
   1. Hypokinesia (Restraint)
   2. Hyperkinesia (Exercise)

E. Disorders induced by Motion

F. Altered Pressure Gradients
   1. Lower Body Negative Pressure

II. *Aquatic Immersion (Buoyancy)*

III. *Hypogravics*

A. Weightlessness

B. Sub-Gravity

IV. *Hypergravics*

A. Rotational Acceleration (centrifugation)
   1. Coriolis Effect

B. Linear Acceleration

V. *Transient Accelerations*

A. Impact (Impulses)

B. Vibration (Harmonics)

Late in 1973 the International Union of Physiological Sciences formed an International Commission on Gravitational Physiology, with Professor Hilding A. Bjurstedt, Sweden, as Chairman. A concern of the Commission is the matter of nomenclature and terminology, and a decision was reached by Professors Bjurstedt, Pace, and Smith to utilize the COSPAR Correspondent Group effort as jointly serving both the COSPAR and IUPS interests.

The objective is to establish a coherent system of units and descriptive words and terms for the rapidly advancing and diversifying science concerned with the biological effects of Earth gravity and other acceleration fields. Of particular interest is the selection of appropriate descriptors that can be used commonly among various languages, or that can be readily and accurately translated.

All interested scientists are invited to participate in this endeavor. Communications may be directed to Professor Arthur H. Smith, Department of Animal Physiology, University of California, Davis, California 95616, U. S. A.

Exobiology

Life Sciences and Space Research XIII — Akademie-Verlag, Berlin 1975

IS THE DETECTION OF OPTICAL ACTIVITY IN EXTRATERRESTRIAL SAMPLES A SAFE INDICATOR FOR LIFE?

W. Thiemann

Institut für physikalische Chemie, Kernforschungsanlage Jülich, FRG

Since Pasteur's times it has been accepted that optical activity and life are intimately correlated with each other, i.e. the one could not exist without the other. This argument logically allows the following speculation related to interplanetary exploration: detection of optical activity in an extraterrestrial sample would be the most straightforward proof of the existence of living matter outside the earth. On the other hand it has been shown that spontaneous production of optical activity can be achieved without the help of living matter, so that the close association of optical activity with life is not necessarily true. The paper discusses the possibilities which allow a proof of some sort of life on other planets by polarimetric methods.

1. Introduction

One of the most exciting scientific challenges for mankind of today is the search for the existence of life outside the planet Earth and its possible detection on one of the neighbour planets within the near future. Several independent methods for the detection of life on Mars have been proposed. The most powerful tools of the Viking mission [1, 2] are the techniques of mass spectrometry in combination with gas chromatography (MS/GC), incorporation of radioactive $^{14}CO_2$ into organic compounds via some sort of photosynthesis-like procedure, and photometry of sterile nutrient broths after incubation with Martian soil. At an earlier stage an application of polarimetric methods for investigating optical activity as an indicator for living material had been discussed, but apparently discarded at some later stage of the planning procedure. The reasons for this decision lay mainly in the ambiguity in the interpretation of results and the very high costs of this experiment [3].

2. The Approach by Polarimetric Measurements

2.1. Optical Activity and Life

Since the days of Pasteur it has been generally assumed that optical activity and life were intimately related, because optical activity (the asymmetry of biological matter) is one of the main characteristics—or even the outstanding feature—of life. This sprang from the discovery that the tartaric acids in biological products are of optically active forms, whereas tedious efforts to synthesize optically active tartaric acids in the laboratory always yielded racemic, i.e. inactive, tartaric acids. The same has been reported in the case of many other chemicals in innumerable experiments. The fact that terrestrial life is based almost exclusively on L-amino acids and D-carbohydrates has always stimulated the thinking of many people. Exceptions from this rule are encountered in the cell walls of certain bacteria and blue-green algae, in many antibiotics, in snake-venoms, and others.

It should be emphasized that in our opinion the borderline between living and non-living is not so easy to define as many people used to think; from this arises the difficulty of looking for a specific reaction or for a chemical structure which could be immediately typified as "biological".

2.2. The Information gained from a Measurement of Optical Activity

One single measurement of an optical rotation at one wavelength—preferably the sodium D line at 589 nm—in an arbitrarily chosen solvent would of course yield almost no information about the origin and the structure of the compound under consideration. It would only show the presence of some optically active compound of undefined structure. The sign and the intensity of the optical rotatory signal depends in a very sensitive way on many parameters: structure of the molecule (position of asymmetric C-atom, type of chromophores, molecular asymmetry, helical molecule, etc.; structure and acidity of solvent [4]; wavelength of measurement; temperature of cell; concentration of optically active compound in solvent; ionic charge; optical purity. Artefacts also occur, so called pseudo-rotation [5], if solutions contain any solid particles or are turbid, due to straying of the incident polarized light beam.

The optical rotation of a defined chemical compound can change sign or increase or decrease in intensity by many orders of magnitude [6], as a result of changing the wavelength of the incident light beam by only about 10 nm or changing the pH by only one unit.

2.3. Technique of Measurement

In spite of all these limitations such qualitative information—optical activity: yes or no—is a rather low cost experiment compared with other life-detecting devices. The measuring device would consist in its simplest set-up mainly of two Nicol prisms, a photomultiplier, and a servo-motor which turned the polarizer into the direction of minimum or maximum energy output of the multiplier. As a light source a simple mercury low-pressure lamp in combination with a few interference filtering glasses would suffice, for this emits many sharp lines in the UV and visible regions of the spectrum and would avoid the relatively expensive monochromator. For ordinary laboratory application Perkin-Elmer sells an instrument [7] that would meet all requirements necessary: it is of relatively low cost, its precision and sensitivity are sufficient (angles of rotation of $\pm 0.002°$ can be recorded reproducibly, if the instrument is adjusted at optimum). I think that such a device in compact form planned for one of the future planetary missions would add only little to the overall costs of an unmanned life-detecting probe to be sent to Mars or other planets of interest.

3. Interpretation of Possible Optical Activity Measurement

3.1. Optical Activity not necessarily restricted to Living Material

The structure of living matter requires asymmetry in its molecular elements. The secondary, tertiary and quaternary structure of proteins or nucleic acids, their helical shape, could not be achieved with perfectly racemic amino acids or racemic carbohydrates. Polymerization of purely racemic monomers results in disordered random coils. The commonly encountered typical α-helix of the proteins [8] is a right-hand screw made up from only L-amino acids. Up to a certain extent on can "cheat" the helical architecture in co-polymerizing a small amount of the D enantiomers, but as one reaches the point of equimolar concentration of D and L (in the monomers) [9], the α-helix becomes a random coil, and changes over to the left-hand screw as one adds D-amino acids in excess of the L-amino acids. The helix-structure of the bio-polymers is found in every living organism as *the* basic design; it appears that optical activity related to the helix-structure is a necessary prerequisite for a biological cell.

On the other hand the complementary hypothesis "biological material is a necessary prerequisite for optical activity", in the sense of Japp [10] is not correct. Logically a single exception suffices to abolish this postulate. A very fine example is the experiment reported by Havinga [11]; it shows unambiguously that *optical*

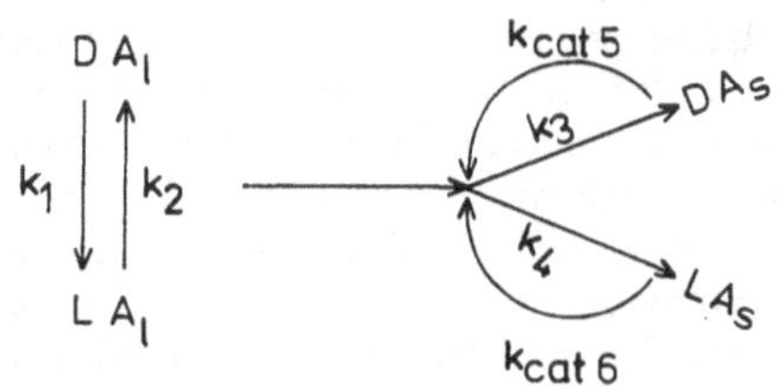

Fig. 1. Principle of Havinga's experiment [11] (after [12]).

activity can be generated spontaneously without the interference of biological matter. Let us think of an optically active compound that racemizes rather fast in solution and rather slowly in the solid phase. Let us further consider a supersaturated solution of the same compound where initiation of crystallization is rather improbable because of the lack of seeds; as soon as a seed for crystallization of the L- or the D-compound is provided precipitation of that enantiomer proceeds rather fast. Calvin [12] depicted this situation in a very simplified way and called it "stereo-specific autocatalysis" (Fig. 1).

If the rate constants $k_1 = k_2$ are very fast (conversion of D-A_l into L-A_l and vice versa in solution) and $k_3 = k_4$ are very slow (precipitation of A_l to A_s), then k_{cat5} is theoretically equal to k_{cat6} and very fast (precipitation after the first seeds have settled) due to the rate being determined by nucleation and crystallization speed. The quaternary ammonium salt methyl-ethyl-allyl-anilinium-iodide in chloroform does meet these requirements. Cooling of a saturated solution below the solubility limit very frequently yields optically active crystals (of D- *or*

L-form), the precipitation of racemic crystals (D- *and* L-forms in equimolar yields) is less probable than the precipitation of optically active crystals (which may conserve their chirality for astronomical times in the solid phase). More recently Pincock et al. [13] described a similar system, the freezing of D- or L-1,1′-binaphthyl from its own racemic melt. There has been disagreement on whether, in spite of the preferred precipitation of D- *or* L-crystals in a single experiment, the overall yield of D to L crystals in a large number of independent experiments would be 1:1. Many authors believe that the average of many experiments would always result in equimolar yields of D and L enantiomers, but we came to the opposite conclusion.

Taking careful precautions against artefacts caused by accidental interference by biogenic material, by physical effects such as optical rotation from quartz cells, from stray light in turbid solutions, by electric drifts of the instrument, by contamination with optically active compounds etc., we were able to "generate" optical activity of one preferred sign by precipitation of racemic sodium ammonium tartrate and asparagine [14, 15] and by polymerization of racemic α-alanine, α-aminobutyric acid, and ε-carbobenzoxylysine into the corresponding polyaminoacids [16]. It has been claimed that optically active compounds could be generated by interaction of circularly polarized ultraviolet light and bremsstrahlung from longitudinally polarized beta-radiation with racemic or prochiral compounds. This hypothesis (originating from Van't Hoff [17] and Byk [18] long ago) was confirmed by Kuhn and coworkers [18] (partial decomposition of azidopropionic acid with left- and right-circularly polarized UV light giving yields of L- or D-azidopropionic acid of about 5% optical purity) and many others [20]. This hypothesis could be very attractive if a stereoselective decomposition or synthesis with circularly polarized bremsstrahlung could ever be experimentally verified: there is a massive abundance of naturally occurring left-polarized beta-radiation on Earth, particularly from ^{40}K and it has been suggested that this influenced the chirality of the biosphere [21]. A convincing proof of a stereospecific interaction of the beta-radiation from radionuclides with racemic compounds has not been presented so far, although there are reported a few interesting experiments which would allow such an interpretation. Among them Garay et al. [22] report a different rate of decomposition of the tyrosine enantiomers in alkaline solution with radiation from ^{90}Sr as well as different life-times of the positron triplet states in the enantiomers of tryptophane in the solid state [23]. However, these experiments have yet not demonstrated the generation of optical activity explicitly. Nevertheless these reports show that there are indeed sources of optical activity other than biological. If this were not the case, a spontaneous appearance of a chiral biosphere from an abiogenic environment $3-4\times10^9$ years ago would have been a mystical event! Even the old "Panspermia" theory of Arrhenius—the infection of the earth with biological seeds from other planets in the universe—has to start with a non-biogenic environment at some point in space and time!

3.2. Detailed Analysis of Optical Activity Pattern

We believe that a careful analysis of optical activity patterns in an extraterrestrial sample would allow a decision on whether it was biogenic or abiogenic in origin in spite of the limitation discussed above. It would not suffice to analyse a single sample for optical activity; one should analyse at least ten or more independent samples. Spontaneous abiogenic precipitation often yields optical activity, as noted, but it is of varying intensity and sign, so this would indicate an inorganic source [11]. If one found instead consistently positive *or* negative rotations, it would be a strong indication for the action of living matter. Living matter is made up of asymmetric chemicals of extremely high optical purity [20], e.g. 100% L-amino acids in the proteins. Abiogenic reactions, such as precipitations or polymerizations, yield substances of low and varying optical purity. Kuhn [19] reported yields of about 5% optical purity. Our own experiments gave up to 15—20% optical enrichments; Havinga's and Pincock's systems [11, 13] yielded up to 80—90% optical purity, but these are rare exceptions to the general rule, and the yields within one experimental series are seldom constant, but vary considerably. The optical purity of biomolecules never varies in a statistical way from one sample to another!

Whereas consistent 100% optical purity would support the probability of the existence of life in extraterrestrial samples, a lower optical purity would not necessarily exclude the existence of life: first there are biological species which contain optically active chemicals of opposite chirality as mentioned above (e.g. D-amino acids in bacteria cells and in blue-green algae) which decreases the overall optical purity of terrestrial biological material; second the death of an organism is, thermodynamically speaking, accompanied by racemization. This was claimed by W. Kuhn [24] and demonstrated by Kvenvolden et al. [25] in a study of the chirality of amino acids deposited in organic sediments. So even an optical purity of less than 100% could still indicate the existence of *formerly living* organisms [26].

In order to determine the optical purity it is of course necessary to know the concentration and the molar rotation of the compound under investigation. This could be performed in an unmanned probe by the built-in chromatographic technique, which yields qualitative and quantitative information of unknown compounds in the sample. The optical rotation has to be continuously recorded at the effluent end of the chromatographic separation column. The most powerful technique for deciding about biogenic or abiogenic origin of optical acitivity would be the study of the structure of the compound from which the optical activity originated. It is very probable that living matter on one of our neighbour planets, if it exists, is similar in biochemistry to that on earth. Amino acids, carbohydrates, nucleic acids, fatty acids, phosphoric esters, etc. would then play the same role as biological elements as on earth. Therefore optical acitivity should be confined to the same groups of chemicals. Optical activity originating, for example, from quartz or potassium silicomolybdate crystals would definitely not indicate the action of life. The safest indicator for life would be the precise analysis of highly optical active amino acids or carbohydrates through a combination of physical and chemical methods. There are strong arguments that life on other planets must make use of optically active amino acids, but there is no cogent reason to postulate that these amino acids will be the same L-enantiomers as those in terrestrial biological material. L-amino acids and D-carbohydrates in living cells on Mars or other planets would support the hypothesis that the parity violation of weak interaction in nuclei [27—30] (with the emission of left-longitudinally polarized electrons) had caused the chirality of the biosphere on earth, while the opposite finding would clearly exclude this hypothesis.

4. Conclusions

Polarimetric measurements would add important information to the search for life on neighbour planets. It would be necessary, however, to perform the following: (i) polarimetry at at least 3 wavelengths of the Hg spectrum to gain

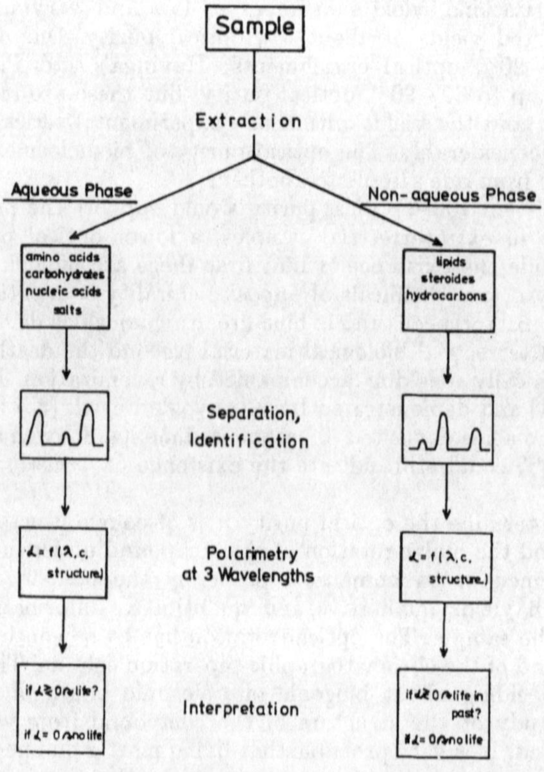

Fig. 2. Scheme of the polarimetric analysis of an extraterrestrial sample.

Table 1
Possible Results of Polarimetric Measurement and their Interpretation

Evidence for the existence of life	Evidence that does not exclude the existence of life
Considerable optical activity of constant sign throughout many samples	Optical activity of statistically varying sign throughout many samples
Optical purity close to 100% (i.e. pure L- or D-enantiomers)	Optical purity varying between 0 and 100% (i.e. mixture of racemic and enantiomer compounds)
Optical activity associated with amino acids, carbohydrates, nucleic acids, and related carbon-containing compounds.	Optical activity of "exotic" nature, associated with non-carbon-containing compounds, such as silicates, heteropolyacids and others.

some information about the optical rotatory *dispersion* of the sample; (ii) polarimetry of many (between 10 and 100) independently collected samples; (iii) determination of the optical purity of the sample; (iv) correlation of optical activity with definite chemical structures.

A general scheme of the analytical technique following extraction (and hydrolysis) which applies the polarimetric method for life detection is depicted in Fig. 2. Table 1 summarizes the possible results and interpretations of a polarimetric measurement of an extraterrestrial sample.

References

[1] H. P. Klein, Origins of Life (formerly Space Life Sciences) **5**, **431** press (1974).
[2] K. Biemann, in: Chemical Evolution and the Origin of Life, North-Holland Publ. Co., Amsterdam 1971 (p. 541); Origins of Life (formerly Space Life Sciences) **5**, 417 (1974).
[3] H. P. Klein, private communication 1973.
[4] W. Thiemann, Tetrahedron **27**, 1465 (1971).
[5] R. Hayatsu, Science **149**, 443 (1965).
[6] W. Thiemann, Tetrahedron **27**, 1470 (1971).
[7] Perkin-Elmer Polarimeter Type 141.
[8] E. R. Blout, P. Doty and J. T. Yang, J. Am. Chem. Soc. **79**, 749 (1957).
[9] P. Doty and R. D. Lundberg, J. Am. Chem. Soc. **78**, 4810 (1956).
[10] F. R. Japp, Nature **58**, 452 (1898).
[11] E. Havinga, Biochem. Biophys. Acta **13**, 171 (1954).
[12] M. Calvin, Chemical Evolution, The Clarendon Press, Oxford 1969 (p. 150).
[13] R. E. Pincock, R. R. Perkins, A. S. Ma and K. R. Wilson, Science **174**, 1018 (1971).
[14] W. Thiemann and K. Wagener, Angew. Chemie (Int. Ed.) **9**, 740 (1970).
[15] W. Thiemann, J. Molec. Evol., **4**, 85 (1974).
[16] W. Thiemann and W. Darge, Origins of Life (formerly Space Life Sciences) **5**, 263 (1974).
[17] J. H. Van't Hoff, Die Lagerung der Atome im Raume, Vieweg, Braunschweig 1894.
[18] A. Byk, Z. phys. Chemie **49**, 641 (1904).
[19] W. Kuhn and E. Braun, Naturwiss. **17**, 227 (1929).
W. Kuhn and E. Knopf, Naturwiss. **18**, 183 (1930).
[20] O. Buchardt, Angew. Chemie **86**, 222 (1974), (including other references!).
[21] A. J. Swallow, Radiation Chemistry of Organic Compounds, Pergamon Press, New York 1960.
[22] A. S. Garay, Nature **219**, 338 (1970).
[23] A. S. Garay, L. Keszthelyi, I. Demeter and P. Hrasko, Chem. Phys. Letters **23**, 549 (1973).
[24] W. Kuhn, Experientia **11**, 429 (1955), (including the older literature on the subject!).
[25] K. A. Kvenvolden, E. Peterson and E. S. Brown, Science **169**, 1079 (1970).
[26] A. I. Oparin, The Chemical Origin of Life, Ch. Thomas Publ., Springfield, Ill., 1964.
[27] C. Sagan, Evolution **XI** (1), 49 (1957).
[28] F. Vester, Seminar at Yale University, 7 Feb. 1957.
[29] Y. Yamagata, J. Theor. Biol. **11**, 495 (1966).
[30] W. Thiemann, Naturwiss. **61**, 476 (1974).

Life Sciences and Space Research XIII — Akademie-Verlag, Berlin 1975

RADIO-CHEMICAL SYNTHESIS OF AMINO ACIDS IN AQUEOUS MEDIA CONTAINING CARBOHYDRATES, HYDROCARBONS AND NITRATES

M. A. Khenokh, E. A. Kuzicheva and E. M. Lapinskaya

Institute of Cytology, Academy of Sciences of the USSR, Leningrad, USSR

The results of ^{137}Cs γ-radiation of aqueous solutions of carbohydrates and hydrocarbons containing nitrates are reported. It was found that aliphatic amino acids were synthesized in solutions containing glucose, mannose, arabinose, benzene, benzoic acid, salicylic acid and octane. Glycine, alanine, leucine, valine, threonine, aspartic acid, lysine and histidine were identified chromatographically.

1. Introduction

In the pre-biological time of our planet chemical evolution of abiogenically synthesized organic substances is thought to have been influenced by various kinds of physical factors, among them, ionizing radiation. The studies here presented, following our previous investigations [1—4], give evidence that amino acid synthesis may have been accomplished by the action of γ-radiation of the primary ocean if it contained carbohydrates, hydrocarbons and nitrate salts.

2. Methods

10^{-1} M solutions of glucose, mannose and arabinose containing 10^{-2} M of nitrate were subjected to γ-radiation (dose-rate 2,64 krad per min.). The source of γ-radiation was ^{137}Cs. The irradiation doses applied varied from 1×10^6 to 26×10^6 rad. A dose of 20×10^6 rad was found to be optimal for the formation of amino acids. 10^{-2} M nitrate solutions saturated with benzene and 10^{-2} M solutions of benzoic and salicylic acids were also irradiated with $12 \times 10^6 — 60 \times 10^6$ rad. The irradiated solutions were freed from salts by passing them through a column packed with Dowex-50 (H^+). Adsorbed amino acids were eluted with 1,5N NH_4OH. Eluates were evaporated and then dissolved in a definite volume of water. An amino acid analyser (KLA-3, Hitachi) was used for identification and quantitative determination of amino acids. Amino acids derived from hydrocarbons were chromatographed on FN-1 chromatographic paper. For quantitative determination spots stained with 0.4% ninhydrin solution in acetone were eluted with 70% ethanol and the amino acid concentration was estimated using calibration curves for glycine, analine and threonine. The following solvents were used: n-butanol, CH_3COOH, H_2O-(4:1:5), 2-butanol, HCOOH, H_2O-(30:15:2), n-propanol, H_2O- (70:30). Experiments performed in the dark revealed no amino acid formation. The experiments were repeated at least three times.

3. Results

In solutions of carbohydrates, radiolysis induced radio-chemical conversions accompanied by the formation of products having characteristic spectra of compounds with λ_{max} 264—269 nm and λ_{min} 238—241 nm. Arabinose, xylose, organic acids and formaldehyde were detected among radiolytic products of glucose. Chromatographs showed four spots that are still not identified. Amino acids occurred in all cases, but they varied in their content and composition. As seen from Table 1 the greatest diversity of amino acids is synthesized when mannose is irradiated in the presence of ammonium nitrate (gly, ala, ser, thr, val, asp, glu, lys, his). The irradiation of glucose and arabinose under analogous conditions yielded the same set of amino acids except valine and glutamic acid.

Table 1

Radio-chemical Synthesis of Amino Acids in Solutions of Carbohydrates containing Nitrates (radiation dose 20×10^6 rad) (t = traces)

Cation	Carbohydrate	Concentration of amino acids $\times 10^{-5}$ molar								
		gly	ala	ser	thr	val	asp	glut	lys	his
NH_4^+	glucose	1.82	t	1.29	t	—	0.06	—	0.44	0.05
	mannose	1.53	0.18	1.01	0.20	t	0.20	t	0.49	t
	arabinose	8.76	4.51	1.11	t	—	t	—	0.47	t
K^+	glucose	0.85	0.74	0.73	—	—	t	—	0.29	0.42
	mannose	0.27	t	0.19	—	t	t	—	t	t
	arabinose	4.28	0.08	0.56	0.02	—	0.06	—	0.45	t
Ni^{++}	glucose	0.69	t	0.26	—	—	t	—	t	t
	mannose	0.89	—	0.38	—	—	t	—	0.08	—
	arabinose	0.75	t	0.42	—	—	0.10	—	t	t

The smallest number of amino acids arose when mannose was irradiated in the presence of Ni^{++} ions. They were gly, ser, asp, lys. These amino acids were synthesized by the irradiation of all three carbohydrates. The glycine yield is higher in arabinose solutions containing NH_4^+ and K^+ ions. Valine was formed only by the irradiation of mannose with NH_4NO_3 and KNO_3. The quantitative yield of amino acids was not large. It is not unlikely that under the experimental conditions some synthesized amino acids were decomposed by the same energy that induced their formation. γ-radiation of nitrate solutions saturated with benzene resulted in the formation of gly, ala, thr, val, leu, lys.

Following irradiation with 18×10^6 rad the largest number of amino acids was formed in the presence of Fe^{+++} ions (Table 2). With increased dosage the synthesis of various amino acids increased, especially in the presence of K^+ and Sr^{++}.

Table 2

Radio-chemical Synthesis of Amino Acids in Nitrate Solutions saturated with Benzene

Cation	Dose $\times 10^6$ rad				
	18	22.7—23.6	41.5—47	54—57	12.6
K^+	gly, ala	gly, ala, thr, lys	gly, ala, (thr), lys	gly, ala, lys	gly, ala
Sr^{++}	gly, ala	gly, ala, val, leu, asp	gly, ala, val, leu, asp	gly, ala, asp	gly, ala
Pb^{++}	gly, ala	gly, ala, lys	gly, ala, lys	gly, ala	—
Fe^{+++}	gly, ala, thr, lys	gly, ala, thr, lys	gly, ala, thr, val, leu	—	gly, ala, thr, lys

Table 3

Radio-chemical Synthesis of Amino Acids in Salicylic Acid Solution containing Nitrates

Cation	Dose $\times 10^6$ rad			
	12.5	29	38	46
K^+	gly, ala, (thr)	gly, ala, thr	gly, ala, thr	gly, ala, thr
Sr^{++}	gly, ala, thr	gly, ala, thr	gly, ala, thr, lys	gly, thr, lys
Pb^{++}	gly (ala)	gly, ala	gly, ala, lys	gly, ala, lys
Ni^{++}	gly	gly, ala	gly, ala, lys	gly, ala, lys

Table 4

Radio-chemical Synthesis of Amino Acids in Octane Solutions in the presence of Nitrates

Cation	Dose $\times 10^6$ rad		
	13	29	41
K^+	gly, ala	gly, ala, thr, val, leu, asp, lys	gly, ala, val, lys
Sr^{++}	gly, ala	gly, ala, thr, val, leu, asp, lys	gly, ala, (thr), (leu), (lys)
Pb^{++}	gly	gly, ala, thr, leu, asp, lys	gly, ala
Ni^{++}	gly, ala	gly, ala, thr, val, leu, asp, lys	gly, ala, thr, (leu)

Further increase of the dose led to a decrease in the number of amino acids and their decomposition. Quantitative determinations of glycine synthesized in the presence of the cation showed that the maximum amount of glycine was formed at a dose of 2.3×10^6 rad. By their effects on the yield of glycine, the cations may be arranged as follows: $Sr^{++} > K^+ > Pb^{++} \geqq Fe^{+++}$. The irradiation of benzoic acid solutions showed the same amino acids as with benzene in the presence of nitrate.

Following irradiation of salicylic acid, the variety of amino acids was less than that with benzene and benzoic acid. Valine, leucine and aspartic acid were absent (Table 3). The table shows that threonine formation occurred only in the presence of K^+ and Sr^{++}. Lysine synthesis was catalysed by Sr^{++}, Pb^{++}, Ni^{++} and absent in the presence of K^+.

The formation of amino acids as a result of γ-radiation in nitrate solutions saturated with octane is shown in Table 4. This shows that at lower radiation doses only glycine and alanine are formed. With increased dosage (29×10^6 rad) the number of synthesized amino acids rose sharply. At still heavier doses part of the amino acids underwent radiolytic decay. The quantitative yield of glycine irradiated with 29×10^6 rad in the presence of K^+ and Sr^{++} was 1.2×10^{-4} mol/l.

4. Discussion

In the epoch preceding the appearance of life on earth abiogenically synthesized substances, carbohydrates and hydrocarbons perhaps among them, probably underwent ionizing radiation in water masses [5]. Nitrates dissolved in the hydrosphere could serve as a source of nitrogen for the formation of amino acids. The synthesis of amino acids from carbohydrates and aromatic hydrocarbons was apparently produced by radio-chemical changes in the aqueous medium. The yield was greatly increased by ammonium cations. By interacting with nitrates they were converted into amino acids with carbon chains of differing lengths. The number of carbon atoms was sometimes consistent with that of the initial carbon chain. Phenol, muconic aldehyde, formaldehyde, diphenyl traces and small amounts of catechol and quinone are formed by radiolysis of benzene in the presence of water and oxygen [6, 7]. It is conceivable that these products of radiolytic changes of benzene can undergo further conversion by the action of ionizing radiation. The carbon-containing fragments and radicals thus arising may have participated in the formation of the carbon chain of amino acids. The tables indicate that alanine and glycine were most resistant to the effect of ionizing radiation in solutions of benzene, salicylic acid and octane. It can be seen that only certain aliphatic amino acids were formed. There is some evidence of a more active radiolytic destruction of amino acids with a long carbon chain. A similar observation has been made on the action of UV light on amino acids [1]. We suggest that polypeptides that presumably were first formed in the primary hydrosphere (upon adsorption of amino acids on clay mineral such as montmorillonite or kaolinite) were composed mainly of amino acids with short carbon chains. The significance of such abiogenic synthesis on clays has been discussed previously [8].

References

[1] M. A. KHENOKH, N. P. BOGDANOVA and E. M. LAPINSKAYA, Problem. Kosm. Biology, **6**, 313 (1971) (in Russian).
[2] E. A. KUSICHEVA and M. A. KHENOKH, Life Sciences and Space Research XI, 67 (1973).
[3] E. M. LAPINSKAYA, M. A. KHENOKH and A. M. SKUPSKAYA, Biochimitcheskaya evolutiya 1973 (p. 21) (in Russian).
[4] M. A. KHENOKH and E. M. LAPINSKAYA, J. Evol. Biochem. Physiol. **10**, 140 (1974) (in Russian).
[5] A. U. OPARIN, Life, its Nature, Origin and Development, Nauka, Moscow 1969.
[6] M. DANIELS, G. SCHOLES and I. WEISS, J. Chem. Soc. 832 (1956).
[7] L. U. KARTASHOWA and A. K. PIKAEV, Chimiya visokih energi **3**, 10 (1969) (in Russian).
[8] E. T. DEGENS and I. NATHEYA, in: Prebiotic and Biochemical Evolution, North-Holland Publ. Co., Amsterdam 1971 (p. 39).

Life Sciences and Space Research XIII — Akademie-Verlag, Berlin 1975

SOME CONSIDERATIONS OF THE THEORETICAL LIMITS FOR LIVING ORGANISMS

P. H. A. SNEATH

MRC Microbial Systematics Unit, University of Leicester, Leicester, England

The study of the limits of life has been traditionally divided into two parts (a) extreme conditions for growth and metabolism, and (b) those conditions permitting survival in a dormant state. The variables most often considered have been temperature, radiation, availability of water and ageing. For growth and metabolism the most sensitive parts of living organisms appear to be those that involve labile metabolites, of which there is a great variety. For dormancy the stability of covalent bonds seems most important; here the C—C bond is likely to be less significant than C—O and C—N. But for both classes of damage the stability greatly depends on numerous factors (such as pH and oxygen concentration) that have seldom been considered in work so far. It is therefore desirable to explore the effects of combinations of adverse factors upon key metabolites and macromolecules. In addition, such studies may permit better estimates of the effects of very long time periods by extrapolation from experimental work.

1. Introduction

The study of the limits of life has usually been considered from two aspects, (a) extreme conditions that permit growth, metabolism and reproduction, and (b) extreme conditions for survival in the dormant state. These correspond roughly to the effects of adverse conditions upon cell metabolism and upon basic structure respectively, and thus provide a convenient framework for theoretical discussions.

The variables most often considered have been temperature, radiation and availability of free water, and there are a number of reviews on these [1—5]. But relatively few studies have been made on combinations of adverse factors except in the field of disinfection, e.g. in relation to sterilization of space craft [6, 7] and the well-known protective effect of dryness upon heat disinfection, which also applies to some higher organisms [8].

One important difference between the aspects (a) and (b) is that in the former the living organism has a greater capacity to protect itself from the environment. This is primarily because the actively metabolizing organism can readily repair damage if it possesses repair mechanisms, whereas a dormant organism has little such capacity, and if it is to survive it must be able to repair damage very quickly when it does come out of the dormant state.

There is a wide spectrum of active and passive resistance to injurious agents. Some, such as certain chemicals, can be excluded by cells by selective permeability barriers or physical ones (e.g. spore walls, encystment), and both active and dormant forms can thus protect themselves. Indeed dormant ones often have the advantage here, as shown by the well-known fact that bacterial spores can survive in 70% ethanol and many other media in which growth is impossible. Others, such as hard radiations and heat, are almost impossible to exclude, at least for small organisms, whether in the active or dormant state, and—in the sense that effects of ageing cannot be prevented by manipulating time itself—time is an agent that cannot be excluded at all. Most agents are intermediate in these respects. Oxygen, acids (low pH), radiation, and so on, are variously excluded by some organisms and not by others. Another form of protection is afforded by detoxification of certain chemicals.

In considering metabolism, the more labile intermediates may well be key targets for environmental damage, though this has been little studied. For survival one would expect the basic structure of the cell (of its proteins and nucleic acids in particular), to be the most important targets, although one cannot exclude the possible significance of reduction of labile metabolites to a level so low that an otherwise undamaged cell structure cannot start to function at all.

2. Limits for Growth and Metabolism

The effect of adverse conditions on growth and metabolism are well reviewed by Skinner [3] and Kushner [5]. Temperatures almost up to the boiling point of water can be tolerated; bacteria can grow in hot springs at 88° and there is circumstantial evidence for growth at over 95° [9]. At the other extreme the limit is the availability of liquid water; micro-organisms can live in brine pools below 0° and the alga *Dunaliella* has been observed active in brine pools at −15° [2, 10]. Substances that protect cells by lowering the freezing point of cell fluids are widespread in Arctic and Antarctic organisms including fishes [11]. It is evident that liquid water is likely to be the major theoretical limit on growth temperatures, especially in view of the experiment recorded long ago by Zobell [12] but never followed up, that under high hydrostatic pressure some bacteria can apparently grow at temperatures of 104°. There seems no evidence that high pressures are a limiting factor for aquatic organisms, because both bacteria and metazoa are active in the abyssal depths [13], and low pressure only affects them by reducing the availability of liquid water.

The extremes of pH tolerance recorded by Skinner [3] and Kushner [5] are from about pH 0.5 for *Thiobacillus thiooxidans* (which oxidizes elemental sulphur to sulphuric acid) to pH 13 for the alkali-tolerant blue-green alga *Plectonema*. However, the internal pH of such organisms appears to be much closer to neutrality, as they are able to regulate this to some extent [4, 5, 14]. Yet there are unusually acid-resistant proteins of *Thiobacillus* outside the ionic barrier, in the form of the flagellar proteins [15]. These are the counterpart of the unusually heat resistant flagellar proteins of the thermophil *Bacillus stearothermophilus* [16], which cannot owe their heat resistance simply to protective mechanisms within the cells. It is also now well established that at least some enzymes of thermophils are more heat resistant than those from their mesophilic counterparts [4].

The availability of energy is of course one critical factor, but it must be noted that numerous organisms can live on very depauperate energy sources by conventional standards. Micro-organisms are able to grow in laboratoy distilled water, wheras cave-dwelling animals often seem to have no obvious source of food. Likewise, some micro-organisms can exploit very small energy gradients: thus the autotrophic bacterium *Thiobacillus ferrooxidans* can use as its sole energy source the small energy potential of oxidizing ferrous to ferric ions [4].

Resistance to dryness is closely bound up with tolerance of low available water, and Kushner [5] discusses briefly the mechanisms involved. Halophils may exclude the sodium ion but contain equal molar amounts of potassium, and many extremely halophilic micro-organisms and metazoa (e.g. brine shrimps) are known. In desert soils there may be a film of liquid on soil particles due to moisture absorbed by hygroscopic salts (commonly $CaCl_2$): this was noted long ago by Darwin in the Peruvian desert near Iquique [17]. Bacterial growth under such conditions, as well as in soils when liquid films are only present during brief periods when moisture is available, is now being studied [18], an aspect of microbiology that greatly interested the late Professor Vishniac. Some fungi are particularly tolerant of dry conditions [19]. Numerous higher organisms have physiological mechanisms for retaining metabolically produced water; probably this is also true of some micro-organisms [20]. Numerous experiments on growth at low humidities make it clear that very small quantities of moisture would permit growth of many terrestrial micro-organisms on Mars [18, 21, 22].

Although numerous micro-organisms [23] are highly tolerant of metal ions (e.g. the fungus *Acontium velatum* growing in saturated cupric sulphate solution, and thiobacilli in mine waters) it is not clear to what extent they simply exclude the metals [4].

Extremely reducing environments are inhabited by anaerobic bacteria but few other organisms. Highly oxidizing environments have been little studied, though some bacteria produce hydrogen peroxide and others produce elemental iodine from iodides, and are highly resistant to iodine [24]. Many micro-organisms are extremely resistant to ionizing radiation both while growing and dormant; they evidently possess very effective repair mechanisms involving sulphydryl groups [25], because reagents that inactivate these make the organisms radiosensitive.

3. Limits for Survival in a Dormant State

There have been few comprehensive reviews of survival in a dormant state since those of Keilin [26, 27]. One or two notable new records may be cited. The most remarkable is perhaps the survival for about 10000 years of *Lupinus* seeds in permanently frozen Arctic subsoil [28]. Bacteria about 1000 years old have survived in lake sediments [29], whereas in soils they seem to survive for only a few hundred years [30]. The earlier reports of survival of bacteria in rock-salt are now attributed to contamination [31], and the shape of the survival curves that would lead rock-salt of several geological epochs to contain just a few viable bacteria per gramme seems most implausible [2]; there is no reason to suppose that all the curves would asymptote at such a level. Another remarkable observation is the great resistance of the dormant desiccated larva of the chironomid fly *Polypedilum vanderplanki* that can survive at 100° if perfectly dry [8]. Bacterial spores of extreme resistance to dry heat are continually being reported [32]. In contrast the report of a protective effect of CO_2 at high pressure on bacteria exposed to moist heat [33] has not been confirmed [34]; though the effective pH is not known, one would expect that under such conditions acid lysis of cell constituents would be likely.

One would expect that survival would be greatest at temperatures close to absolute zero, as suggested by Becquerel [35], and many micro-organisms and some higher organisms (e.g. the desiccated larva of *Polypedilum vanderplanki*) can survive cooling to such low temperatures. The extrapolation of existing knowledge of survival times of spores of *Bacillus subtilis* [2] to very low temperatures would (if the curve followed approximately the Arrhenius equation) imply survival for at least 10^6 years, and probably 10^9 or more. The limit may be set by the accumulation of lethal mutations. Drake [36] notes that the tendency for mutations to be proportional to time rather than to number of generations suggests that dormant genomes naturally accumulate mutations, though we do not know to what extent we are observing the failure of repair mechanisms. We have little information on the effect of low temperatures on mutations. The estimated temperature dependences vary widely [36], which suggests that some are due to endogenous chemical mutagens or copying errors, and are hence temperature dependent.

Presumably ionizing radiation could cause mutations in cells at absolute zero, and if so the radioactivity of the environment becomes important over long time periods. At very long periods the damage due to decay of the potassium isotope ^{40}K within a spore could yield a dose of about 750 rad in 10^9 years [30], which would kill only a small proportion of spores; however, if the recoil energy had pronounced biological effects then a much higher proportion would be killed. Yet for times between 10^6 and 10^9 years there seems no theoretical reason why some spores should not survive at absolute zero, and in principle, therefore, dormant organisms could survive in meteorites.

4. Stability of Metabolites

Denaturation of enzymes, damage to cell membranes and mutations in nucleic acids are probably the major targets on which adverse factors act, but before discussing these we may note that there are numerous labile active metabolites. Many of these are high-energy compounds like ATP, and being natural energy donors they would seem likely to break down rather readily. This is doubtless so, and in this connection it has been pointed out [4] that ATP is naturally acid-labile, and that this may set a limit to the lowest pH that a cell can tolerate. There is some justice in this argument, but is should be noted that such high-energy compounds as ATP, NADP, acetylcoenzyme A, etc., must also be very stable. They need, like catalysts, to have an inherent stability of the bulk of their molecules, coupled with instability in a particular portion, the part that can store chemical energy (for example in phosphate residues). Only in this way can the molecules have the reversible property of taking up and giving out energy without becoming irreversibly broken down. I feel, therefore, that there need be no theoretical need for the sensitive target compounds to be necessarily labile ones. It is of course true that if all the high-energy compounds become "discharged" the cell may not be able to re-form them.

5. Stability of Structural Molecules

The mode of damage to the basic structural molecules of the cell—nucleic acids and proteins—is especially significant for longevity in a dormant state because the cell cannot resume metabolism if its main structure is destroyed. There is, however, evidence that heat and ionizing radiation may act mainly on proteins and nucleic acids respectively. Radiation can certainly be a very lethal agent, but even small amounts of physical protection can shield organisms effectively.

Thus there are likely to be numerous local habitats where living systems could persist and evolve, even if ultraviolet radiation or cosmic rays reached the earth's surface in very large amount (e.g. ocean depths, sand bars, caves).

Stability to heat is of wider significance. We have seen that microbial spores from soil can resist dry heat at temperatures close to charring, so that the stability of the common organic covalent bonds may well be the limit to their heat resistance. The three bonds that seem to be most important in structural molecules are C—C, C—N, and C—O. These have bond energies in kcal $mole^{-1}$ of 81.1, 77.9 and 88.2 respectively as calculated in the conventional manner [37]. The backbone of a protein chain consists of alternate C—C and C—N bonds, with the side chains of the amino acids attached through another C—C bond. The ketonic bond C=O and the amino bond N—H that alternate with the side chains are much more stable (180 and 102 kcal $mole^{-1}$ respectively) whereas the other bonds of importance in proteins and nucleic acids (C—H, O—H, and P—O) are also very stable. One would expect that the most heat-labile bond in a peptide chain would therefore be the C—N bond, so that at temperatures close to charring many of these bonds would be soon broken. However, the stability of a bond of a molecule depends also upon the adjacent molecules, and in particular the activating effect of the ketone group on adjacent bonds suggests that the ether C—O bonds in the sugar moieties of nucleic acids may be more liable to damage by heat than the C—C bond. The resulting aldehyde groups are reactive, leading to further breakdown. This agrees with the familiar observation that carbohydrates and proteins char at lower temperatures than paraffin wax.

However, the side chains of the different amino acids are not all equally stable. Abelson [38, 39] made a number of experiments on the relative stability of amino acids in aqueous solution, and noted that some were very unstable. We can envisage, therefore, that certain amino acid residues could decompose at a lower temperature than the peptide links, yielding inactive enzymes, for example. Thus phenylalanine could be hydroxylated to tyrosine [40]. This is directly relevant to stability of proteins over long periods of time.

Another suggestion arises from the lability of DNA to acid. The most sensitive change appears to be the deamination of cytosine residues to uracil, and Drake [36] has suggested that low pH coupled with high temperatures may greatly accelerate mutation by increasing the corresponding transition $G{:}C \rightarrow A{:}T$. However, we do not know what the pH is inside cells such as *Sulfolobus* that can grow at high temperatures and also low pH [41], so we cannot be sure whether this sets any theoretical limit on growth in hot acidic habitats. It is of interest, however, that Pullman and Pullman [42] have calculated the resonance energy of purines and pyrimidines, and find that the pyrimidines are the less stable, which is in broad agreement with Drake's observations on heat-induced mutation, though Pullman and Pullman find thymine should be slightly less stable than cytosine, an effect not noted by Drake.

The heat stability of lipids has been discussed by Brock [4] in relation to thermophily, because there is some evidence that the cell membrane of thermophilic bacteria may be unusually stable, and Brock notes that there may be some necessary property of the organelles of eukaryotes (such as a loose structure) that limits the possible heat resistance of eukaryote organisms.

The above considerations bear on whether it is possible for biological macromolecules to persistunchanged over long periods of time. This has been recently reviewed at length by Florkin [43]. Until the observation that mollusc shell protein from the Eocene still gave a biuret reaction [44] it was thought that proteins would persist only for brief periods. Florkin and his colleagues have now evidence of somewhat altered shell protein in material even of Palaeozoic age, and of collagen from dinosaur eggs of Cretaceous date. The exact nature of the alteration with time is still not clear; the amounts of amino acid decrease and the amount of bound ammonia increases with age, so there is probably some breaking of the peptide chains as well as decomposition of the side chains. However, there are evidently long peptide molecules still present, as the decalcified material resists dialysis, and electron microscopy shows the expected microstructure. Some amino acids (phenylalanine, tyrosine, cysteine, methionine, tryptophan) seem to be absent in the fossil material, or almost so, whereas others increase in relative concentration. Very similar reports by others have been reviewed by Ferdinand et al. [45].

Chitin seems to be very stable, persisting almost unchanged from the Cambrian [46]. There is some evidence that lipids do not persist very long in fossils [45, 47] but this has been little studied, nor has much been done with nucleic acids, though their general absence suggests they are yet more unstable.

6. Discussion

Despite the considerable body of work that has been done in the field, very little attention has been paid to the way the local environment may modify the effects of adverse conditions on biological materials. This is especially notable in the work on stability of molecules. The theoretical bond energies and the like may be very poor indicators of what actually happens, because of the profound effect of the local conditions. Thus in the presence of oxygen, or at low pH, the stabilities cited above may be entirely misleading. Many examples could be given: the stability of some spores to heat seems dependent largely on exclusion of oxygen; tryptophan is well known to be very labile at acid pH, but it is quite stable at pH 9 [40]; different experimental conditions in a Fischer-Tropsch type of process with added ammonia could account for the presence or absence of compounds like porphyrins in ancient deposits [48, 49]. Morowitz [50] has shown how one can simulate on a computer certain factors such as these.

Local factors such as Eh, temperature and pH could have pronounced effects on the persistance of biological materials, and explain some discrepancies. Perhaps it is to this that we should look to explain the difference between Abelson's pioneering work and later studies cited by Florkin [43]. Abelson [38, 39] found that serine, threonine, arginine and tyrosine were very unstable in fossil shells, whereas recent work suggests that all except tyrosine are very stable [43, 45, 46].

Similarly there is evidence in the banded iron formations of the upper Onverwacht that local concentrations of oxygen may have occurred even though there was little or none in the atmosphere until much later [51], and this could affect the interpretation of hydrocarbon composition in these very early sediments [52]. It is clear, too that the pH of rocks varies widely from acidic sulphide ores to alkaline limestones.

As Florkin [43] has noted it is now becoming important to study experimentally the conversion of protein, and other materials of biological interest, to "paleoprotein" and its analogues. The stability of amino acids may be different in free and combined form. Also the actual products of amino acid breakdown might be studied: thus a tryptophan residue in a peptide might be converted to a serine, for example. Among the major problems in investigating the origin of organic materials in rocks is to know whether the experimental work done to support the studies does realistically mimic the effects of very long time periods at much milder conditions, and a better knowledge of local factors would be of great help here.

As to the effects of adverse factors on metabolism, there have been scattered reports in the literature that especially rich culture media are needed for the recovery of damaged micro-organisms [53], such as frozen or freeze-dried bacteria. This may reflect the loss of labile metabolites that must then be provided exogenously. This has never been studied in detail, but it would certainly repay attention.

References

[1] J. R. Vallentyne, Ann. N. Y. Acad. Sci. **108**, 342 (1963).
[2] P. H. A. Sneath, Discovery **25**, 20 (1964).
[3] F. A. Skinner, Proc. Roy. Soc. B **171**, 77 (1968).
[4] T. D. Brock, Symp. Soc. Gen. Microbiol. **19**, 15 (1969).
[5] D. J. Kushner, in: Chemical Evolution and the Origin of Life, North-Holland Publ. Co. Amsterdam 1971 (p. 485).
[6] S. Kaye, in: COSPAR Technique Manual No. 4, 133 (1968).
[7] L. O. Kallings, in: COSPAR Technique Manual No. 4, 157 (1968).
[8] H. E. Hinton, Nature **188**, 336 (1960).
[9] T. L. Bott and T. D. Brock, Science **164**, 1411 (1969).
[10] G. H. Meyer et al., Science **138**, 1103 (1962).
[11] R. E. Feeney and R. Hofman, Nature **243**, 357 (1973).
[12] C. E. ZoBell, Producer's Monthly **22**, 12 (1958).
[13] R. Y. Morita, Oceanogr. Marine Biol. Ann. Rev. **5**, 187 (1967).
[14] D. J. Kushner, Exper. Chemotherapy **2**, 113 (1964).
[15] R. N. Doetsch et al., Antonie van Leeuwenhoek **33**, 196 (1967).
[16] H. Koffler, Bact. Rev. **21**, 227 (1957).
[17] C. Darwin, Geological Observations on South America, Smith, Elder & Co., London 1846.
[18] W. V. Vishniac and S. E. Mainzer, Life Sciences and Space Research XI, 25 (1973).
[19] W. J. Scott, Adv. Food Res. **7**, 83 (1957).
[20] S. I. Aksenov et al., Life Sciences and Space Research XI, 55 (1973).
[21] E. Packer et al., Icarus **2**, 293 (1963).
[22] A. A. Imshenetsky et al., Life Sciences and Space Research XI, 63 (1973).
[23] R. L. Starkey and S. A. Waksman, J. Bact. **45**, 509 (1943).
[24] R. S. Gozlan and P. Margalith, J. Appl. Bact. **36**, 407 (1973).
[25] M. J. Thornley, J. Appl. Bact. **26**, 334 (1963).

[26] D. KEILIN, Science Progress in the Twentieth Century **41**, 577 (1953).
[27] D. KEILIN, Proc. Roy. Soc. B **150**, 149 (1959).
[28] A. E. PORSILD et al., Science **158**, 113 (1967).
[29] T. CROSS, personal communication.
[30] P. H. A. SNEATH, Nature **195**, 643 (1962).
[31] J. DE LEY et al., Antonie van Leeuwenhoek **32**, 315 (1966).
[32] W. W. BOND et al., Appl. Microbiol. **20**, 573 (1970).
[33] P. MOLTON et al., Nature **243**, 242 (1973).
[34] K. A. SOUZA and L. P. ZILL, Nature **247**, 67 (1974).
[35] P. BECQUEREL, C. R. Acad. Sci. Paris **231**, 1274 (1950).
[36] J. W. DRAKE, Symp. Soc. Gen. Microbiol. **24**, 41 (1974).
[37] M. J. S. DEWAR, The Electronic Theory of Organic Chemistry, Clarendon Press, Oxford 1949.
[38] P. H. ABELSON, Ann. N. Y. Acad. Sci. **69**, 276 (1957).
[39] P. H. ABELSON, Proc. V Int. Congr. Biochem. **3**, 52 (1963).
[40] N. FRIEDMAN et al., in: Chemical Evolution and the Origin of Life, North-Holland Publ. Co., Amsterdam 1971 (p. 123).
[41] J. L. MOSSER et al., Science **179**, 1323 (1973).
[42] B. PULLMAN and A. PULLMAN, Quantum Biochemistry, Wiley Interscience, New York 1963.
[43] M. FLORKIN, in: Chemical Evolution and the Origin of Life, North-Holland Publ. Co., Amsterdam 1971 (p. 10).
[44] M. FLORKIN et al., C. R. Acad. Sci. Paris **252**, 440 (1961).
[45] W. FERDINAND et al., Comp. Biochem. Physiol. **44B**, 889 (1973).
[46] D. B. CARLISLE, Biochem. J. **90**, 1c (1964).
[47] S. K. DAS et al., Comp. Biochem. Physiol. **23**, 519 (1967).
[48] M. CALVIN, Trans. Leicester Lit. Phil. Soc. **62**, 45 (1968).
[49] C. C. YANG and J. ORÓ, in: Chemical Evolution and the Origin of Life, North-Holland Publ. Co., Amsterdam 1971 (p. 155).
[50] H. J. MOROWITZ, in: Chemical Evolution and the Origin of Life, North-Holland Publ. Co., Amsterdam 1971 (p. 37).
[51] P. CLOUD, Amer. Scientist **62**, 54 (1974).
[52] W. M. SCOTT et al., Nature **225**, 1129 (1970).
[53] J. R. POSTGATE and J. R. HUNTER, J. Appl. Bact. **26**, 405 (1963).

Life Sciences and Space Research XIII — Akademie-Verlag, Berlin 1975

MEMBRANE DAMAGE IN DEHYDRATED BACTERIA AND ITS REPAIR

M. Frankenberg-Schwager, G. Turcu, C. Thomas, H. Wollenhaupt and H. Bücker

Arbeitsgruppe biophysikalische Raumforschung, University of Frankfurt, Frankfurt, FRG

The dehydration of bacteria by vacuum exposure results in damage to the cell membrane. This membrane damage does not necessarily lead to cell death. A part of the dehydrated bacteria is capable of eliminating the membrane damage by repair processes. Repair can proceed rapidly under conditions that permit synthesizing activities. The kinetics of this repair process were studied by means of the membrane-mediated biosynthesis of the cell wall as well as by the recovery of resistance to small concentrations of lysozyme. Repair is a precondition for cell proliferation. At low temperature cells can conserve their membrane damage and the repair process can be initiated when conditions become favourable.

In previous investigations the effect of vacuum on micro-organisms was studied in simulation chambers and in space [1—6]. This paper deals with the effect of dehydration on the survival of bacteria and the mechanism on which this effect is based. The bacteria used were *Escherichia coli*. Dehydration was achieved by exposing the bacterial population to a vacuum of about 10^{-7} torr [7]. This results in the inactivation of a certain fraction of the bacteria. Moreover, the dehydrated cells show an increase in the membrane permeability as indicated by the release of cell components absorbing UV light of about 260 nm wavelength (Fig. 1) [8]. The inactivation is correlated with the increase in the membrane permeability of dehydrated bacteria (Fig. 2). This finding indicates that the dehydration of bacteria results in damage to the cell membrane. This statement is confirmed by another experimental approach. Bacteria surviving dehydration show an abnormal behaviour in buffers which are commonly used for the storage of bacteria in suspension; in contrast to the control bacteria, bacteria surviving dehydration die during incubation in these media. This can be interpreted in two ways: either the dehydration of bacteria leads to the inability of the cell membrane to retain its function of osmotic control; alternatively, dehydration leads to a defect in the bacterial cell wall while leaving the cell membrane intact. In the latter case the destroyed cell wall would no longer inhibit the expansion of the cell volume by the uptake of water and the cells will finally explode. This second possibility was ruled out, because bacteria surviving dehydration could not be kept alive in hypertonic media which allow survival of bacteria with an intact cell membrane whether or not their cell wall is defective.

The results of these two experiments, increase in permeability and the loss of osmotic control, together indicate that the dehydration of bacteria leads to damage of the cell membrane.

Now two questions arise: (i) is the inactivation of only a fraction of dehydrated bacteria caused by membrane damage? (ii) can bacteria surviving dehydration repair their membrane damage? The first indication of a repair process came from the comparison of the growth curves of bacteria surviving dehydration and

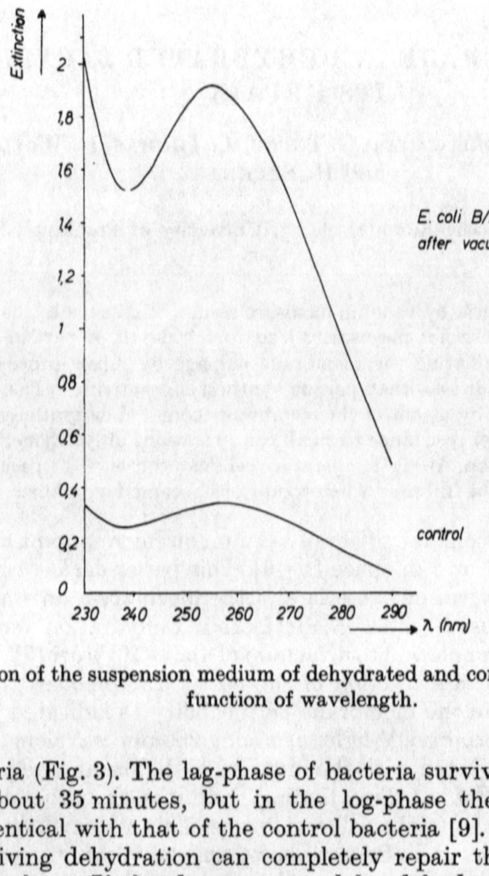

Fig. 1. Extinction of the suspension medium of dehydrated and control cells of *E. coli B/r* as a function of wavelength.

control bacteria (Fig. 3). The lag-phase of bacteria surviving dehydration is prolonged for about 35 minutes, but in the log-phase the growth rate of these bacteria is identical with that of the control bacteria [9]. It seems therefore that bacteria surviving dehydration can completely repair their damage within the prolonged lag-phase. If the damage caused by dehydration can be repaired it should be possible to prove this by repair kinetics. The kinetics of the repair of membrane damage can be studied by membrane-mediated biosynthesis of the cell wall. This synthesis was followed by the inactivation of cell wall synthesizing capacity in bacteria by penicillin. This effect of penicillin is based upon its inhibition of the peptide cross connections in the growing murein structure of the cell wall. Cells which do not synthesize murein are therefore not affected by penicillin. Fig. 4 shows the result of this study. Control cells start with the synthesis of their cell wall within a few minutes after incubation in nutrient medium. This is shown by their early inactivation by penicillin. However, bacteria surviving dehydration show a delay of about 32 minutes in the beginning of their cell wall synthesis. Moreover, the initial rate of synthesis is reduced compared with that of the control cells. Only towards the end of the lag-phase does the rate of synthesis of the cell wall equal that of the control cells.

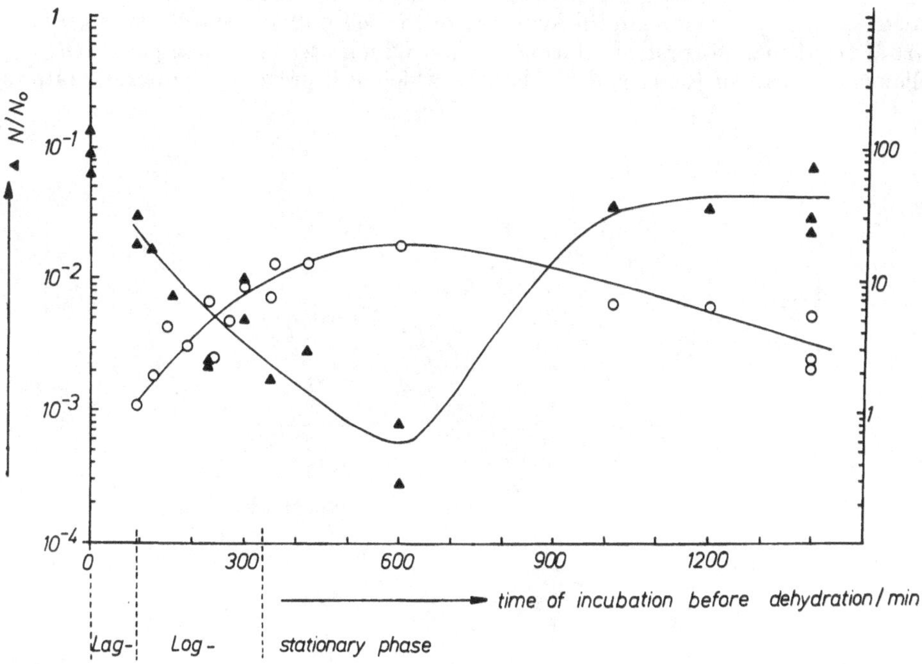

Fig. 2. Correlation of the dehydration-induced inactivation of bacteria (N/N_0) (▲) and the change in the permeability (P/P_0) of the bacterial cell membrane (○) as a function of the bacterial growth phase.

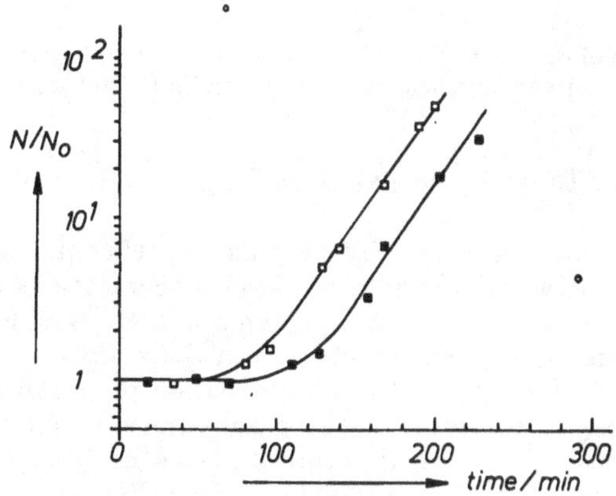

Fig. 3. Growth (N/N_0) of dehydrated bacteria (■) and control bacteria (□) in nutrient broth as a function of incubation time.

From this experiment the following conclusions can be drawn, as far as the functions of the membrane in the synthesis of the cell wall are concerned: a great part of the damage is repaired during the first 32 minutes of the lag-phase; then follows the repair of the remaining damage which still prevents the normal rate

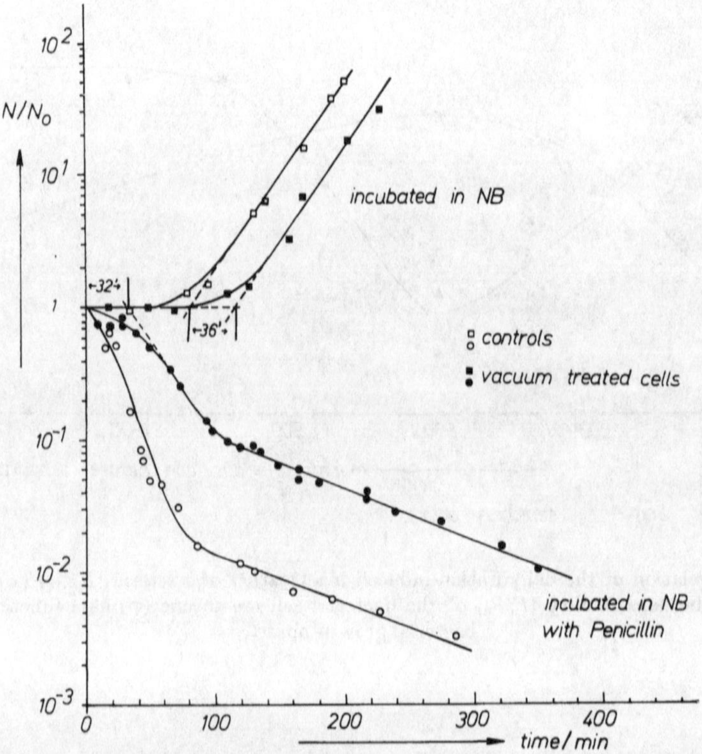

Fig. 4. Growth (N/N_0) of dehydrated (solid symbols) and control bacteria (open symbols) in nutrient broth NB (■, □) and nutrient broth + penicillin G (200 units/ml) (●, ○) at 37 °C.

of cell wall synthesis. Towards the end of the lag-phase the repair process is completed.

These conclusions are confirmed by another experiment. Bacteria surviving dehydration show a reduced resistance to those concentrations of lysozyme that do not affect the control cells at all. Lysozyme is a cell-wall-attacking enzyme which splits bonds in the murein structure, and their repair is mediated by the cell membrane. Fig. 5 shows that after an incubation period in nutrient medium for about 35 minutes the bacteria surviving dehydration have recovered almost completely their resistance to small concentrations of lysozyme. In addition, repair seems to proceed continuously within the first 35 minutes of the lag-phase, because the resistance is only partially restored after the first 15 minutes of incubation. These two pieces of evidence indicate that membrane damage induced by dehydration can be eliminated by a repair process. The repair takes place mainly in the first 35 minutes of incubation in nutrient medium and may be the cause of the prolonged lag-phase. The completion of repair is shown by the restoring of the normal rate of cell wall synthesis, and recovery of resistance to small concentrations of lysozyme. The conditions required for a successful repair process turned out to be a sufficient energy source, a well-balanced ionic environ-

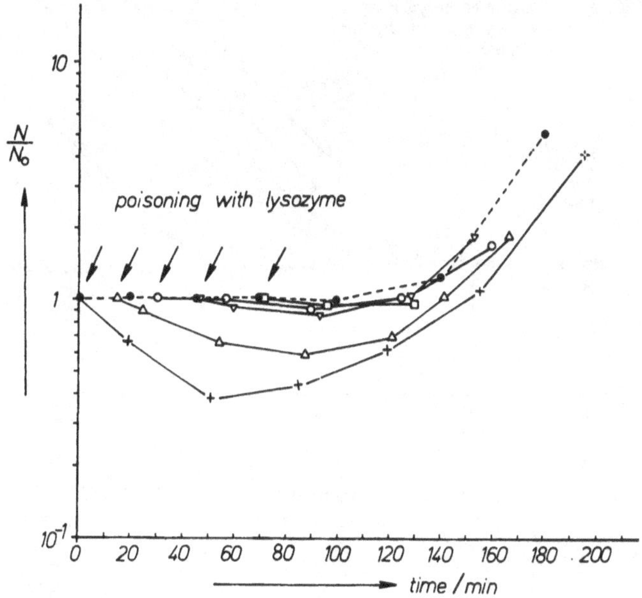

Fig. 5. Response of dehydrated bacteria incubated in nutrient broth to the addition of lysozyme (25 μg/ml) at different times during the lag-phase. Control cells are incubated in nutrient broth only (dashed curve). The arrows indicate the time when lysozyme was added. N/N_0 is defined as relative number of bacteria.

ment and an optimal temperature of 37 °C. These requirements indicate that the repair is connected with synthesizing activity of the cells.

If bacteria surviving dehydration are kept at a low temperature (4 °C) for several hours the membrane damage is not repaired nor does this lead to cell death (Fig. 6). This can be concluded from the fact that the storage of these bacteria at a low temperature does not influence the growth pattern of the bacteria that survive dehydration. Their lag-period is exactly the same whether or not the cells have been kept at a low temperature beforehand. This means that at low temperatures cells surviving dehydration can conserve their membrane damage, and that repair processes can be initiated when conditions become favourable.

On the one hand a correlation has been shown between inactivation of bacteria and an increase of bacterial cell membrane damage due to dehydration. On the other hand bacteria surviving dehydration show repairable membrane damage. From these two facts one can conclude that the membrane damage induced by dehydration is at least one of the causes of dehydration-induced death of bacteria.

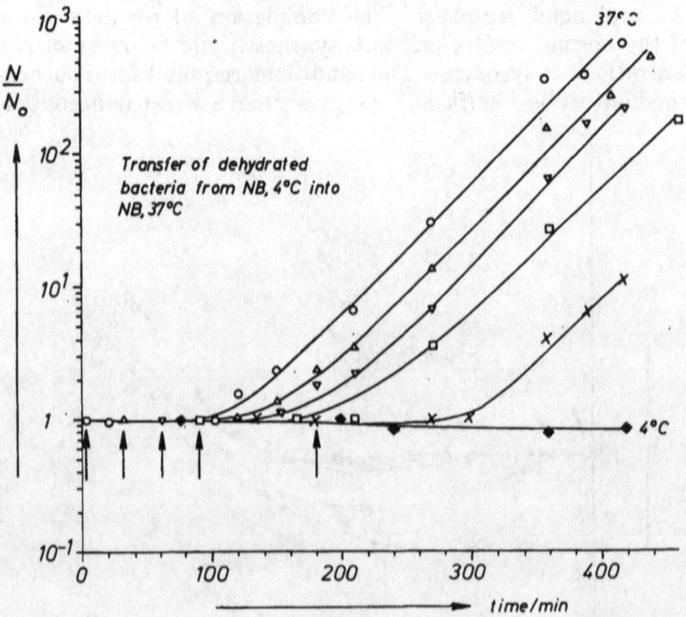

Fig. 6. Transfer of dehydrated bacteria from nutrient broth NB at 4°C to nutrient broth at 37°C. The arrows indicate the time when bacteria were transferred from 4°C to 37°C.

References

[1] H. Bücker and G. Horneck, Biophysik **6**, 69 (1969).

[2] H. Bücker and G. Horneck, Life Sciences and Space Research VIII, 33 (1970).

[3] G. Horneck, H. Bücker and H. Wollenhaupt, Life Sciences and Space Research IX, 119 (1971).

[4] H. Bücker, G. Horneck, R. Facius, M. Schwager, C. Thomas, G. Turcu and H. Wollenhaupt, Life Sciences and Space Research X, 191 (1972).

[5] H. Bücker, G. Horneck, H. Wollenhaupt, M. Schwager and G. R. Taylor, Life Sciences and Space Research XII, 209 (1974).

[6] H. Bücker, G. Horneck and H. Wollenhaupt, in: Proc. Symp. on Microbial Response to Space Environment, NASA Tech. Mem. X-58103, 1973 (p. 21).

[7] M. Schwager, Space Life Sciences **4**, 271 (1973).

[8] H. Bücker, G. Horneck, G. Turcu and C. Thomas, Archives Roumaines de Pathologie Experimentale et de Microbiologie **31**, 1, 7 (1972).

[9] H. Bücker, G. Horneck and H. Wollenhaupt, Biophysik **7**, 217 (1971).

Planetary Quarantine

Life Sciences and Space Research XIII — Akademie-Verlag, Berlin 1975

FROM MARS WITH LOVE

R. S. YOUNG[a] and D. L. DEVINCENZI[b]

[a]Planetary Biology Program, NASA Headquarters, Washington, D.C., USA
[c]Life Sciences Division, NASA Ames Research Center, Moffett Field, Calif., USA

It is possible to return surface material from Mars to the earth. Such a mission is highly desirable from a scientific point of view. Returning samples directly to the earth may not be acceptable from the point of view of back-contamination without sample sterilization; however, sample sterilization decreases the total scientific value, depending on the measurements to be made and the technique of sterilization. Inorganic chemistry is least affected by heat sterilization and biology and organic chemistry are most affected. Other forms of sterilization, e.g. radiation or chemical treatment, seem to be more destructive to one or the other of the scientific objectives than heat, making heat the apparent method of choice at this time. Quarantine of Mars samples in earth orbit seems very difficult and expensive. Containment and subsequent quarantine of unsterile samples returned directly to earth seems difficult, but feasible if proper precautions are taken and facilities provided.

Soil samples can now be returned to earth from another planet using unmanned spacecraft. One possible consequence of this ability is the potential for returning a viable extraterrestrial organism that may interact with terrestrial life forms.

The bioscience community is polarized on the issue of back-contamination of earth; some feel the risk is virtually nonexistent while others believe the risk is high. There is no question that scientific interest in exploring the surface material of another planet is great. The problem arises in determining the method of study that is the most productive and cost effective, that is, *in situ* on the planetary surface with automated landers or by direct study of returned samples in terrestrial laboratories.

Recent studies indicate that unmanned MSSR missions of various types are technically feasible and could utilize much existing hardware and technology.

Probably the most desirable option for studying extraterrestrial material is to return a carefully protected sample to earth for study in the laboratory. For mission simplicity and lower cost, a direct return to earth is optimal. From the point of view of science, a sample that has not been sterilized or altered severely in any way would also be most desirable. The first problem is to determine our ability to return an unsterilized sample "safely". First, we must be assured that the probability of the return vehicle crashing on earth would be very low. While the question is extremely difficult to analyze from the viewpoint of back-contamination, such an analysis is critical. For example, if the chance of crashing and thereby releasing the sample to the earth is only one in a million (a purely arbitrary figure for this argument), it may prove to be an acceptable risk. If, on the other hand, the chance of crashing is substantially higher than one in a million, it may be unacceptable. In addition, we must calculate the risk involved in removing the sample and its container (presumably hermetically sealed) from the spacecraft and conveying it safely to the laboratory where a biological barrier system has been designed, behind which the sample can be safely analyzed.

One MSSR mission option might be to return a heat-sterilized sample to a manned or unmanned orbiting laboratory or earth-based laboratory. The question then arises as to the worth of a sterilized Martian soil sample to the biological and geological communities in view of the scientific objectives of returned sample analysis.

By definition, the active biological processes of growth, reproduction, motility, irritability, and metabolism would be destroyed by heat sterilization. Depending on the sterilization conditions used, however, other meaningful biological measurements might be made on the sample. For example, some biological structure might be preserved and recognized. Certain classes of biochemical could be detected and analyzed, with special attention paid to characteristics such as molecular structure and optical activity. However, only inferences could be made regarding the existence of life on Mars, and the experimental approaches would be limited to chemical and optical techniques. No active biology measurements would be possible.

A sterilized Mars soil sample would still retain much of its scientific value to physical scientists. However, if heat sterilization is used, the maximum temperature reached, as well as the kinetics, is crucial. For example, most of the standard radiometric rock age-dating methods would be unaffected at temperatures below 250 °C. However, problems could be anticipated with argon diffusion between 300° and 400 °C, depending on the mineral composition and size of individual crystals or mineral grains in rock fragments. For the types of samples considered most probable (i.e. igneous rock fragments and mineral grains, etc.), major and minor element and mineral analyses should not be affected if the samples are heated to only a few hundred degrees. However, water, sulfur, and other volatiles could be lost, depending on exactly how they are bound in the sample. The effect of heat sterilization on the organic content of Martian soil can also be predicted somewhat. Based on information obtained from thermal release and pyrolysis experiments on geological samples, it appears that heating samples at 200 °C and below will volatilize some organic compounds, but will not destroy or totally alter all the organic material. Under these conditions, much information can be retained for organic geochemical and exobiological purposes, especially if the volatile products are trapped and later examined. Trapping volatilized organics (on silica gel or Tenax) for later terrestrial analysis should be practical. Combustion of volatilized species, after which the resultant carbon dioxide is trapped, will at least preserve isotopic information. If this process is done as a function of volatilization temperature, the results might be somewhat informative.

Heating the sample to 400 °C for prolonged periods of time will volatilize and decompose most of the organic compounds of significance to exobiology. In summary:

1. Sample return missions from Mars are feasible in the 1980's.

2. The least expensive missions (direct sample return without sterilization) may be criticizable from a back-contamination point of view, although upgrading the handling and containment facilities could make unsterile return acceptable.

3. Sample sterilization decreases the total scientific value appreciably, depending on the measurements to be made. Geology is least affected and biology and organic chemistry are most affected.

4. Quarantine in earth orbit, in the same sense as for the lunar samples, would not be feasible without very large increases in cost. Designing the orbital quarantine facilities, either automated or manned, would be very expensive, risky, and of limited use because of size limitations.

5. Orbital quarantine may be feasible if the sample is split, part of the sample sterilized and returned to earth for study and the remainder studied for pathogenicity in the automated mode as best we can in the limited space available in orbit. Ground studies of sterilized material plus "live" studies in orbit may convince us of the safety of returning the remaining sample to earth under carefully prescribed conditions.

6. Additional unmanned, Viking-type missions to Mars can add considerably to our knowledge about a Martian biota, or its absence, and thus increase the likelihood of being able to return safely an unaltered sample to earth.

The full manuscript of this paper is published in *Science* **186**, 495 (1974).

Life Sciences and Space Research XIII — Akademie-Verlag, Berlin 1975

SELF STERILIZATION OF BODIES DURING OUTER PLANET ENTRY

A. R. Hoffman, W. Jaworski and D. M. Taylor

Jet Propulsion Laboratory, California Institute of Technology, Pasadena, Calif., USA

As a body encounters the atmosphere of an outer planet, whether accidentally or by plan, it will be subjected to heat loads which could result in high temperature conditions that render terrestrial organisms on or within the body non-viable. To determine whether an irregularly shaped entering body, consisting of several different materials, would be sterilized during inadvertent entry at high velocity, the thermal response of a typical outer planet spacecraft instrument was studied. The results indicate that the Teflon-insulated cable and electronic circuit boards may not experience sterilizing temperatures during a Jupiter, Saturn, or Titan entry. Another conclusion of the study is that small plastic particles entering Saturn from outer space have wider survival corridors than do those at Jupiter.

1. Introduction

Many researchers [1—3] have recognized the difficulty in designing a Jupiter entry probe that would survive to the regions of interest of the Jovian atmosphere. Because of this, many were, and still are, of the opinion that it would be extremely unlikely for an unsterilized flyby or orbiter spacecraft to survive (and thereby contaminate) the Jovian atmosphere during inadvertent entry.

To justify this assertion analytically, the Pioneer project office at the NASA Ames Research Center (ARC) performed a study to determine whether spacecraft components inadvertently entering the Jovian atmosphere at high velocity could survive [4, 5]. To summarize the published ARC conclusions: (i) all metal pieces will be consumed during entry; (ii) plastic pieces capable of ablation may survive entry, but such pieces must have characteristic sizes of 0.2 to 0.5 m at the start of entry to survive; (iii) biological contamination by an unsterilized spacecraft impacting Jupiter is "highly doubtful".

A previous COSPAR report [6] presented the work done on the thermal response of small particles, insulating blankets, antenna, and supporting struts when the spacecraft passes through the free molecular regime of the atmospheric entry. It was determined that the spacecraft will break up as a result of heating at the boundary of the continuum flow regime, whereupon the antenna, struts, and blanket will disintegrate and burn away. The only survivors will probably be the science platform and bus, which will enter the continuum flow regime. In this paper, the thermal response of the science platform and part of the bus are discussed, and representative test results are presented. In addition, the extended analysis of small particles, which also covers entries into the atmosphere of Saturn, is included.

2. Technical Discussion

The major part of the effort to be reported is concerned with determining the thermal response of the science platform. A typical ultraviolet spectrometer (UVS), a very complex instrument mounted on the platform, was selected for a detailed thermal response analysis largely because of the abundance of plastic

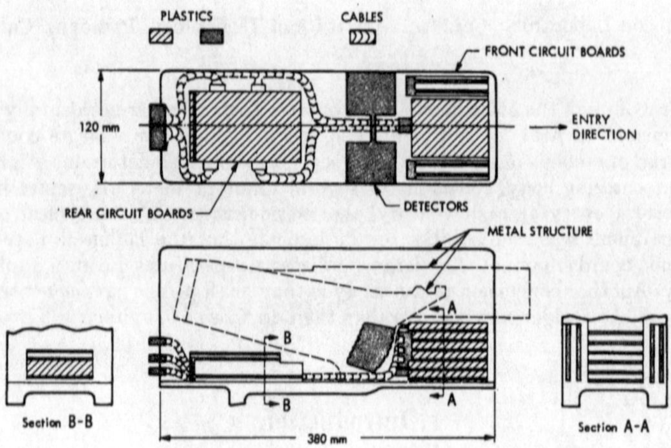

Fig. 1. Location of plastic materials in ultraviolet spectrometer.

materials concentrated in the lower part of the instrument. Fig. 1 schematically illustrates the location of these materials.The plastic materials found within the instrument were G10 epoxy-fiberglass, Castall, Diallyl phthalate, polycarbonate, and Teflon. Decomposition data on these materials were generated by testing the samples in a precision-type thermogravinometer, where each sample was decomposed in a vacuum under constantly increasing heat and the loss of the sample weight was measured as a function of temperature.

In addition to the decomposition data, the thermophysical properties of the above materials as well as their chemical make-up were compiled or otherwise obtained by tests. The materials were also subjected to baking at 650°C in a specially designed vacuum-type oven. The latter was necessary to provide information on the disintegration status of the materials at the melting temperature of aluminium, which constitutes a primary structure of the instrument.

The first item subjected to the analysis was the UVS cable, consisting of copper wires insulated with Teflon. A typical cable cross-section is depicted in Fig. 2. The cable is approximately 13 mm in diameter. The copper wires, Teflon jackets, and voids between the cables are clearly shown. It was assumed for the purposes of the analysis that the cable length was about 400 mm, and that its configuration was straight. This approach, although conservative with regard to the UVS instrument, was found to be justified when cable configurations in the spacecraft bus were studied. Conservatively, entry along the cable axis was assumed. The two cases analyzed (considered extremes) were −15° entry into the warm atmo-

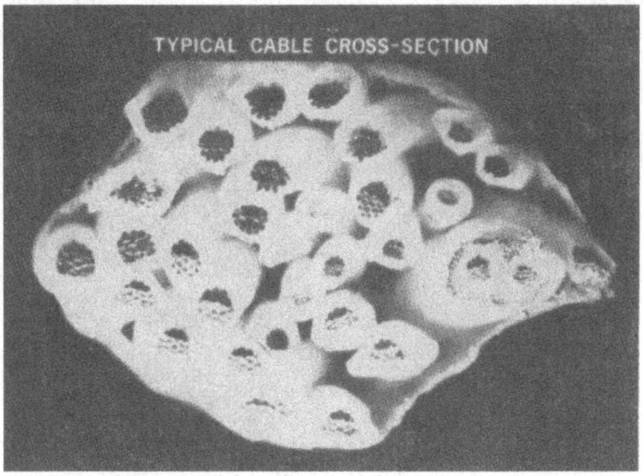

Fig. 2. Typical cable cross section, ~13 mm diameter.

sphere and −90° entry into the cool atmosphere, as defined by the engineering models for Jupiter [7]. It was assumed initially that the Teflon insulation could be considered separately, and that the influence of the copper wires could be superimposed on the results obtained.

Fig. 3 presents the analytical results of Teflon decomposition for the warm atmosphere entries. It is seen that only a part of the 400-mm-long insulation will be disintegrated, and that the remaining length will not develop a temperature

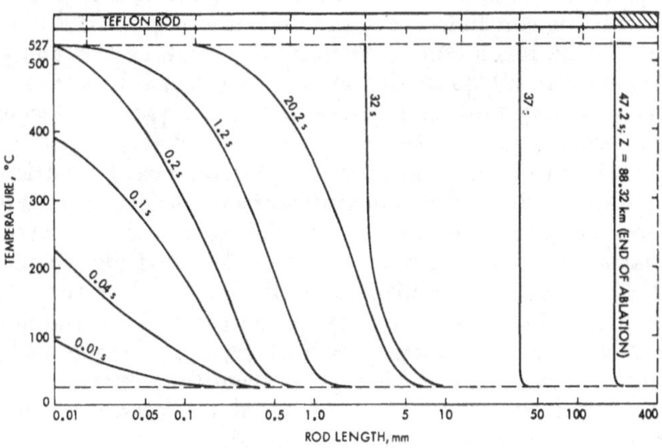

Fig. 3. Temperature profiles of a Teflon rod for Jupiter entry at −15° entry angle: warm atmosphere, initial altitude 750 km.

level high enough to sterilize all the undecomposed portions of the cable. The temperature responses for the simulated entry were calculated down to cloud level in order to establish the insulation status at that altitude.

To account for the effects of wire heating and melting on decomposition of the insulation, an ablation test was performed on a 240-mm-long sample of the cable. An oxyacetylene torch was used to ablate the wire and the insulation. The flame was moved from right to left at a rate corresponding to a complete ablation of

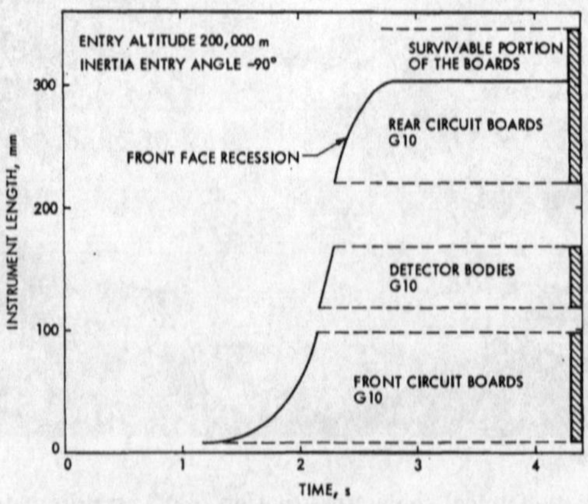

Fig. 4. Ultraviolet spectrometer disintegration of plastics due to heat of Jupiter entry: entry at −90°, cool atmosphere.

the cross section of the cable (Teflon and wires combined). Dual thermocouples were installed at each indicated location on the center wire: one on the bare wire and the other on the outside of the insulation. It was found that approximately 3 mm from the flame front, the temperatures of the wire and the insulation were between 500 and 600 °C (i.e. at the maximum temperature of Teflon decomposition). This is a relatively small temperature difference between the wires and insulation, in spite of the proximity to the melting wires, where the temperature was 1080 °C. It can be concluded that the presence of wires did not significantly affect the process of insulation decomposition, lending the analytical results described previously an adequate degrec of validity. An inspection of the spacecraft bus reveals that typical cable harnesses are frequently thicker and longer than the sample analyzed.

The next items collectively analyzed were circuit boards (cards) and the detector bodies, both made of G10 epoxy-fiberglass. Referring to Fig. 1, and assuming that entry into the atmosphere is to the right of the depicted instrument, we can see that a set of front circuit boards is followed by a bank of detectors and, finally, a set of the rear circuit boards. Each set will, in turn, be exposed to the heat pulse upon disintegration of its predecessor. It was assumed that the thin aluminium cover would instantly melt away, and that the rest of the aluminium structure, placed some distance away from the concentration of plastics, would also melt and be blown away ahead of the plastic disintegration. The melting of the aluminium very close to or adjacent to the plastic materials will be delayed by the heat "blockage" caused by the great amount of gas resulting from plastic decomposition. In short, the adjacent metal will be melting at approximately the speed of plastic decomposition, ensuring some structural integrity during the process.

Fig. 4 presents the results of the analysis. In the case of $-90°$ entry into the cool atmosphere, the third group (rear circuit boards) will be partially decomposed. To evaluate in-depth temperature profiles for $-90°$ cool atmosphere entry, two profiles made at the end of recession were compared. It can be concluded from this comparison that in the $-90°$ cool atmosphere entry, the temperature rise will be less, leaving the unablated portion of rear circuit boards unsterilized. Small particle heating calculations were done for two areas: (i) the effect of the increased heat transfer rate at low speed in free-molecular entry; (ii) the contraction of the small size, small entry angle, non-sterilizing corridor for particles liberated at successively greater levels of ambient atmospheric density. Fig. 5 shows features of a typical small particle entry into the Jupiter warm atmosphere.

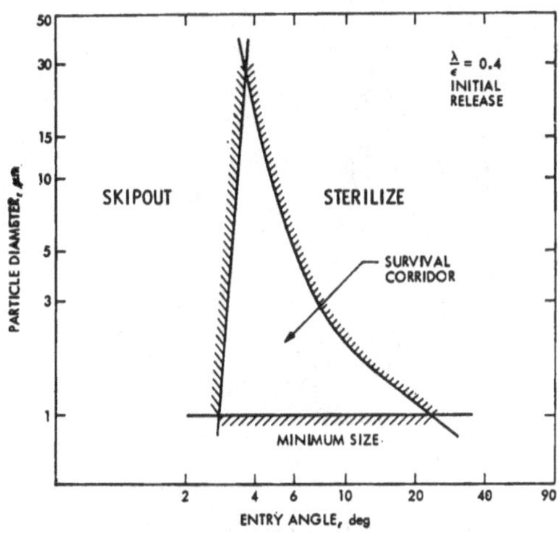

Fig. 5. Survival corridor for small particle entry: Jupiter nominal atmosphere.

Fig. 6 illustrates the contraction of the survival entry corridor for Jupiter, which moves progressively toward smaller entry angles and smaller particle sizes with an increase of the atmospheric density at which the particles leave the spacecraft. Similar calculations were performed for the other Jupiter atmospheres and for the three Saturn model atmospheres. Figs. 7 and 8 show these results for a Saturn and a Titan calculation, respectively. In addition, a preliminary analysis has been done to determine the entry body and trajectory conditions likely to result in the least total heating for the major components of a spacecraft entering Saturn. Total convective plus radiative heating in the stagnation region, per unit mass, was calculated for bodies of different size and weight, in entry at $-90°$, $-15°$, and $-3°$ into the three Saturn model atmospheres. The main differences in the case of Saturn are: (i) the heating is always considerably less than in the Jupiter case when analogous entry heating parameters are considered; (ii) the radiative heating, which comes into play at large nose radius for heavy bodies, is a smaller fraction of the total heating, i.e. radiative plus convective; (iii) the total heating at −15° entry is considerably less than at −90° and −3°. This is in contrast to the Jupiter case, where the result of a high heating rate for a short entry time at −90° entry is comparable with that of a low heating rate for a long entry time at −3°, as well as that of the −15° entry case.

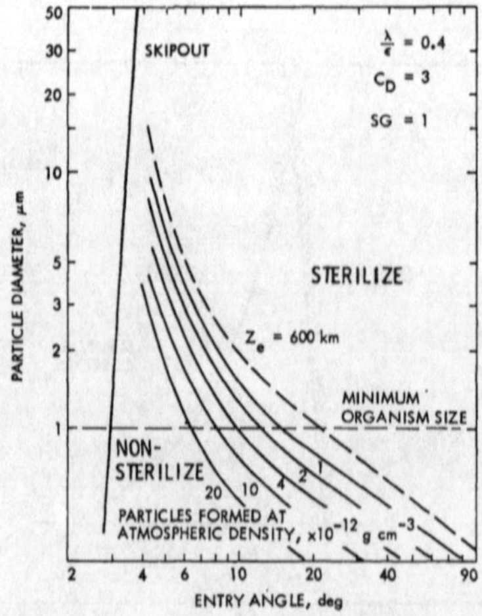

Fig. 6. Effect of delayed release on small particle survival corridors: Jupiter nominal atmosphere. λ is accommodation coefficient, ε is emissivity, c_D is drag coefficient and SG is specific gravity.

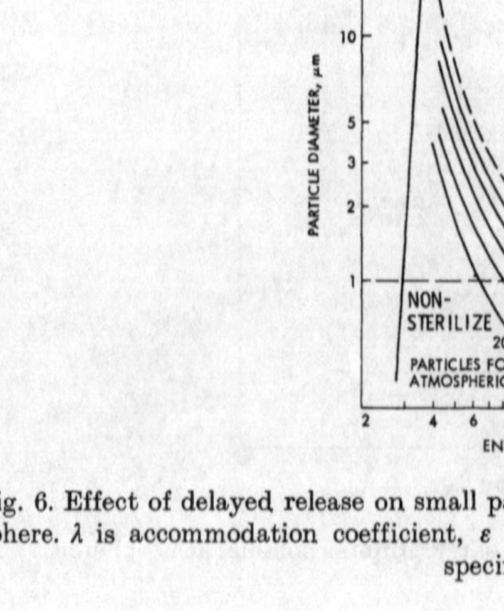

Fig. 7. Effect of delayed release on small particle survival corridors: Saturn nominal atmosphere.

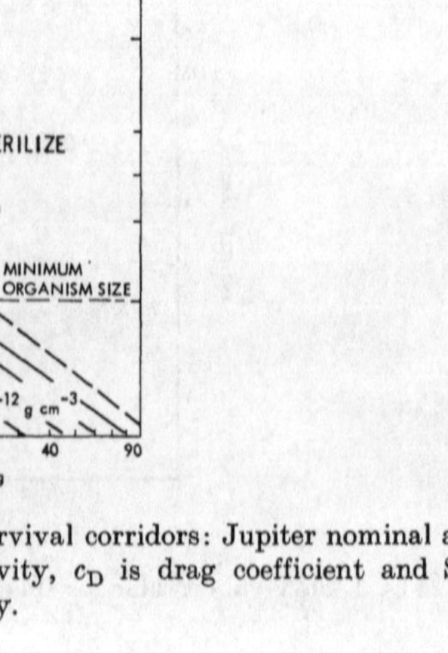

Fig. 8. Effect of delayed release on small particle corridors: Titan nominal atmosphere.

3. Conclusions

3.1. Jupiter

Large Impactables. In contrast to previous expectations, a possibility of survival of organisms which reside within the large impactables (spacecraft, launch vehicles, or large pieces thereof) has been uncovered in the present analysis. The surviving remnants of plastic materials may descend down to the cloud level, where environmental conditions of growth are regarded as favorable. Among the most critical components are the cable harnesses, which have a finite chance of surviving Jupiter entries without being sterilized. This is also true of circuit boards where microbial survival depends on the grouping, size, and orientation of the boards; those placed in a series stand a better chance of surviving. It has been shown that the last group of boards in the ultraviolet spectrometer may not reach sterilizing temperatures during a $-90°$ entry into the cool atmosphere. The results presented are also applicable to the elements of the spacecraft bus, which have a design configuration similar to that of the UVS.

Small Particles. Small particles, which include spheres, flakes, and needles with characteristic dimensions in the range of 1 to at least several hundred μm, have little likelihood of unsterilized entry into Jupiter. The exception is small particles entrained with a spacecraft at first entry, for which a narrow non-sterilizing entry corridor exists in a shallow entry. The entry corridor, ranging in size from 1 to approximately 30 μm, rapidly contracts if the particle liberation is delayed until ambient atmospheric density levels in the range of 1 to 20×10^{-12} g cm^{-3} are reached. The corridor vanishes when liberation occurs at ambient atmospheric density levels above about 30×10^{-12} g cm^{-3}. Particles generated later from spacecraft breakup (mainly around a density level of about 10^{-9} g cm^{-3}) have no chance of unsterilized entry.

3.2. Saturn

Large Impactables. Based on a cursory study of large impactable entry into the atmosphere of Saturn, it is concluded that the components with the best chance of survival in Saturn entry are the heavy components with large nose radius entering at an angle of about $-15°$.

Small Particles. The same general conclusions are obtained concerning survival of small particles during Saturn entry as for Jupiter entry except for displacement of entry corridors. The corridor for a particle vanishes when liberation occurs at ambient atmospheric densities of 50×10^{-12} g cm^{-3}.

3.3. Titan

Large Impactables. Although no detailed studies have been performed, it is believed that unsterilized bodies entering the atmosphere of Titan are more likely to have more components surviving the entry heating than in the case of Jupiter or Saturn.

Small Particles. The survival corridor for small particles is wider than that for Saturn. The corridor vanishes when liberation occurs at ambient densities of $\sim 10^{-11}$ g cm^{-3}.

4. Acknowledgments

The small-particle entry results were calculated by A. McRonald, and the overall advanced missions task was coordinated by C. Gonzalez, both of the Jet Propulsion Laboratory. The cooperation and support of these individuals is deeply appreciated and gratefully acknowledged.

This paper represents the result of one phase of research carried out at the Jet Propulsion Laboratory, California Institute of Technology, under Contract No. NAS 7-100, sponsored by the National Aeronautics and Space Administration.

References

[1] W. VULLIET, S. SAND and W. FINDLEY, Final Report, JPL Contract 952480, Gulf General Atomic (30 June 1969).
[2] W. JAWORSKI, Jet Propulsion Laboratory Space Programs Summary, 37–58 vol. III, p. 146 (31 August 1969).
[3] M. TAUBER and R. WAKEFIELD, J. Spacecraft and Rockets 8, 630 (1971).
[4] B. SWENSON, Rep. TM X-2276, NASA, Washington D.C. (April 1971).
[5] B. SWENSON, Body Shape Effects upon Survivability during Jovian Entry, Ames Research Center, Calif. (1972).
[6] C. GONZALEZ, W. JAWORSKI, A. MCRONALD and A. HOFFMAN, Life Sciences and Space Research XII, 221 (1974).
[7] The Planet Jupiter, NASA SP-8069 (1970).

Life Sciences and Space Research XIII — Akademie-Verlag, Berlin 1975

NEW METHODOLOGY FOR ASSESSING THE PROBABILITY OF CONTAMINATING MARS

D. W. North, B. R. Judd and J. P. Pezier

Stanford Research Institute, Menlo Park, Calif., USA

New methodology is proposed to assess the probability that the planet Mars will be contaminated by terrestrial micro-organisms aboard a spacecraft. Present NASA methods are based on the Sagan-Coleman formula, which states that the probability of contamination is the product of the expected microbial release and a probability of growth. The proposed new methodology extends the Sagan-Coleman approach to permit utilization of detailed information on microbial characteristis, the lethality of release and transport mechanisms, and of other information about the Martian environment. Three different types of microbial release are distinguished, and for each release mechanism a probability of growth is computed. Using this new methodology an assessment has been carried out for the 1975 Viking landings on Mars. The resulting probability of contamination for each Viking lander, 6×10^{-6}, is based on expert judgment and is amenable to revision as additional information becomes available.

1. Introduction

The biological contamination of Mars is a complex issue involving a great variety of scientific, engineering, and policy considerations. In many areas the information available is limited. Nonetheless, NASA is committed to a planning process derived from the 1966 COSPAR resolution that is based on assessment of the probability of planetary contamination. The task facing the authors in a recent project for NASA was to criticize the existing assessment methodology, develop appropriate new methodology, and apply the methodology to the 1975 Viking lander [1]. While the application deserves further review from the scientific community, the results indicate a probability of contamination well below the mission constraint of 10^{-4} imposed by NASA. We believe that the new methodology represents a significant advance over present NASA practice in assessing the probability of planetary contamination.

2. Review of Sagan-Coleman Approach

The probabilistic model of planetary contamination advanced by Sagan and Coleman [2] was the stimulus and theoretical foundation for the 1966 COSPAR resolution. Their procedure used the approximation

$$P(C) = E(N)\, P(G),$$

where C = the event that Mars will be contaminated by terrestrial organisms aboard the spacecraft, N = the number of viable terrestrial organisms (VTOs) released to the Martian environment or into its atmosphere from the spacecraft (a random variable),

$E(N) = \sum_{k=0}^{\infty} k\, P(N = k)$, the expected (or mean) number of VTOs released, G = the event that a single VTO will grow, meaning that it would survive and multiply in the Martian environment.

This formula forms the basis for the assessments of the probability of planetary contamination as they are currently carried out by NASA.

The most serious problem in the use of the Sagan-Coleman formula is that $P(G)$ and $E(N)$ have been used to refer to a randomly selected organism, with no specification of the type of organism or how and where it is introduced into the Martian environment. This approach ignores important available information and places an exceedingly difficult task on the scientific experts who are asked to assign $P(G)$. By a relatively straightforward extension of the Sagan-Coleman approach, the problem can be overcome.

3. Distinguishing among Organisms

To survive and reproduce on Mars a microbe should be facultatively anaerobic, that is, able to reproduce in the absence of oxygen. All terrestrial life requires water in a usable form. Water usable by microbes is unlikely to exist on Mars at temperatures significantly above 0 °C, so the microbe should be facultatively psychrophilic, that is, able to reproduce in a temperature range of 0 °C or below. For our analysis we assumed that only organisms that are facultatively both anaerobic and psychrophilic contribute significantly to the probability of contamination. The importance of organism characteristics was noted six years ago by Sagan et al. [3].

The UV radiation flux on the Martian surface is strong enough to kill any unprotected terrestrial micro-organisms in minutes. A micro-organism implanted directly into Martian soil will have a far better chance of surviving than a microbe that rests for many days on the exterior surface of the spacecraft. We therefore distinguished three mechanisms for release of organisms into the Martian environment: (i) direct *implantation* of a microbe into Martian soil, (ii) release by aeolian *erosion* presumably during a Martian dust storm, and (iii) release from the surface of the spacecraft into the Martian atmosphere due to mechanical *vibration*, thermal effects, or other means. Dependence on landing site was not included in our analysis, but if future information indicated that the availability of usable water on Mars varies significantly with the release location, the model could be expanded to include this dependence.

4. New Assessment Methodology

An overview of the model for assessing the probability of planetary contamination is shown in Fig. 1. The model is composed of four submodels that describe successively the bio-burden on the Viking Lander, microbial release mechanisms,

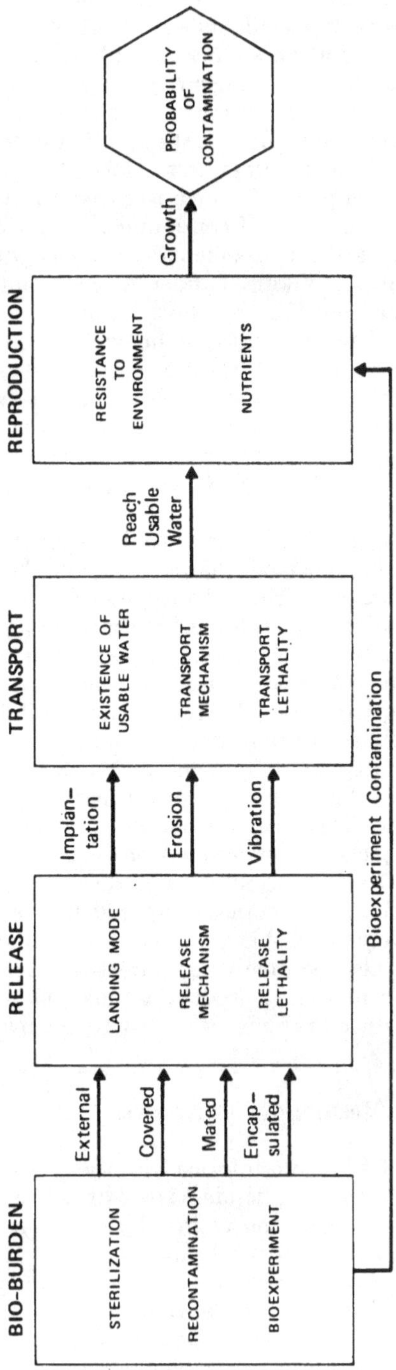

Arrows represent transfer of viable terrestrial organisms (VTOs).

Fig. 1. Mission contamination model.

transport in the Martian environment, and the resistance of terrestrial microbes to the Martian environment and the availability of nutrients needed for microbial reproduction on Mars. Communication among the submodels is through the expected number of VTOs that undergo various specific events, such as release from the spacecraft. The overall output from the model is the expected number of organisms that reproduce on the planet. Contamination is defined as reproduction on the planet by one or more organisms. Since the expected number of organisms that reproduce on the planet is much less than unity, we can interpret this output quantity as the probability of contamination of Mars.

The first submodel provides the subsequent submodels with the number of micro-organisms existing on the Viking Lander when it lands on Mars. This biological load is characterized not only by the type of micro-organism but also by its location on the lander. Included in this submodel are the number and type of organisms at various locations prior to sterilization, the reduction in bio-burden effected by the sterilization requirements, possible recontamination, and increase or decrease of the micro-organism population during transit to Mars. Assessment of the probability of contaminating Mars from contamination of the bioexperiments on the lander is calculated directly; the contribution from this term is not significant.

The Release submodel uses the bio-burden profile as input. It represents explicitly the uncertainty in the landing mode (hard or soft) and the release mechanism (implantation, erosion, or vibration). The lethality of each mechanism is assessed, and the number of VTOs released by each mechanism is calculated.

Unless a microbe from the lander is directly implanted in a hospitable micro-environment providing water, Martian winds or other transport mechanisms are needed to transport it there. However, the microbe may be killed or immobilized by UV radiation or other causes during transit. These transport and lethality processes have been represented by a dynamic probabilistic model, specifically a Markov process. Each of the three release mechanisms corresponds to a separate starting state in this process. The output from the Transport submodel is the expected number of VTOs reaching a micro-environment with usable water.

Finally, given that a VTO has reached a hospitable micro-environment with usable water, we examine the circumstances required for its reproduction. The organism must be facultatively anaerobic and able to reproduce at temperatures near or below 0 °C. It must also be able to acquire the nutrients necessary for reproduction. The output from the Reproduction submodel is the number of organisms expected to reproduce in the Martian environment.

5. Results of the Analysis

Given the present state of scientific information, the probability of biological contamination by each of the two Viking landers is 6×10^{-6}. This value is approximately a factor of 16 below the mission constraint of 1×10^{-4} imposed by NASA.

Fig. 2, which reproduces the structure of the model presented in Fig. 1, indicates the crucial variables and the major intermediate results at each point in the model. The expected number of VTOs transferred from one submodel to the next is indicated on each arrow linking the components. Also, the box representing each submodel contains a list of the critical variables pertaining to that part of the model.

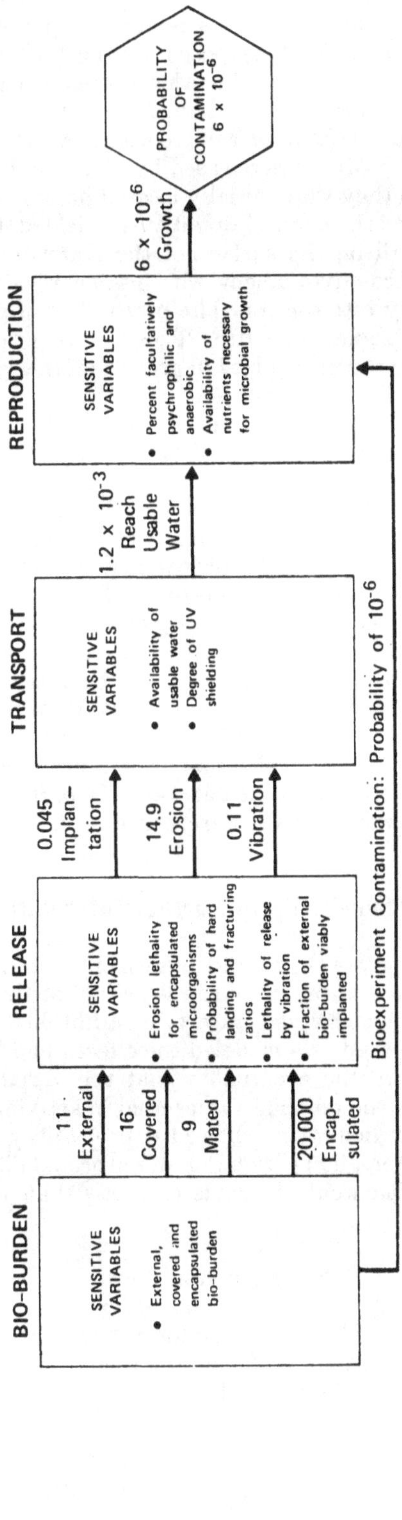

Numbers on arrows give the expected numbers of VTOs.

Fig. 2. Mission contamination model results.

Another important result, not apparent in Fig. 2, is that the probability of growth of a microbe varies widely with its release mechanism. A VTO released by implantation is not immediately exposed to the UV radiation and has a probability of growth of 2.8×10^{-5}. At the other extreme, a microbe released by erosion must survive transport in a Martian dust storm and is 100 times less likely to grow and reproduce than a microbe released by implantation; its probability of growth is 2.8×10^{-7}. Microbes released by vibration have an intermediate chance of surviving. Since they were initially located on exposed surfaces and are released in a viable state, they must already be shielded from UV radiation. However, because they fall on the surface of the Martian soil, they have less chance of reaching a micro-environment with usable water than do microbes that are implanted directly into the soil. The probability of growth for microbes released by vibration is about 5.3×10^{-6}. These findings clearly indicate the importance of conditioning the probability of growth on the release mechanism.

6. Insights from Sensitivity Analysis

The above results reflect the present state of scientific information, which is characterized by large uncertainties. The sensitive variables, for which the uncertainty has a significant effect on the probability of contamination, are given in Fig. 2. The probability of contamination is especially sensitive to two characteristics of the Transport submodel: the probability of finding water on Mars and the lethality of transport. The two variables of the Reproduction submodel are also critical. If growth is to occur the microbe must be resistant to the Martian environment, and is must find appropriate nutrients. Experimentation in microbiological laboratories could address the question of whether different microorganisms surviving the dry heat sterilization cycle could reproduce in a Martian micro-environment if supplied with usable water and UV protection. Unfortunately, little attention has been given to this research until recently [4].

7. Overall Sensitivity to Further Information

To determine the sensitivity of the overall assessment to additional information, an approximate calculation was performed in which 16 of the input variables in the assessment model were considered uncertain. Additional information would cause these inputs to be revised. We modeled the effect of additional information by assigning probabilities to the eventuality that the variables would take on higher or lower values than the nominal values used in arriving at the assessment of 6×10^{-6}. The resulting calculations showed a probability of a few per cent that the constraint of 10^{-4} might be violated and a probability of 50% that the probability of contamination would be revised to less than 10^{-6} on the basis of the additional information.

8. Conclusion

We have shown how available scientific information can be structured into a model for assessing the probability of contamination. There is currently less confidence in the probabilities input to the model than in the structure of the model itself. We believe that the use of a formal model as a basis for planning quarantine policy represents a substantial advance over NASA's current approach, which relies on the single parameter $P(G)$. By providing a detailed structural basis for assessing the probability of microbial growth, the model facilitates critical review and revision by the community of scientists concerned with planetary quarantine.

We have recommended to NASA that the current procedure of determining mission sterilization requirements on the basis of a single probability of growth be replaced by a procedure that distinguishes among types of organisms, types of release mechanisms, and other characteristics that affect whether an organism released from a spacecraft will reproduce in the environment of another planet.

Important decisions will be made in the coming years about missions to the outer planets and the return of a surface sample from Mars. It would be highly desirable to have a comprehensive decision framework to address the question of quarantine strategy. The suitability of decision analysis concepts to the quarantine problem has already been pointed out [2]; the methodology and procedures have been applied to similar complex problems in space project planning [5—7], and other large-scale scientific research programs [8]. The formulation should make explicit the interaction between quarantine procedures and spacecraft cost and reliability, value judgments about microbial proliferation, and the goals of the planetary exploration program.

References

[1] B. R. Judd, D. W. North and J. P. Pezier, Final Rep., SRI Project MSU-2788, NASA Contract NASW-2535, Stanford Research Institute (1974).

[2] C. Sagan and S. Coleman, Astronaut. Aeronaut. **3**, 22 (1965).

[3] C. Sagan, E. C. Levinthal and J. Lederberg, Science, **159**, 1191 (1968).

[4] T. L. Foster, Rep. No. 3, Hardin-Simmons University (1973).

[5] J. E. Matheson and W. J. Roths, paper presented at Nat. Symp., Saturn/Apollo and Beyond, American Astronautical Society, 11 June 1967.

[6] J. E. Matheson and A. B. Pollard, Final Rep., SRI Project 6152, Stanford Research Institute (1967).

[7] M. M. Menke and B. R. Judd, Final Rep., SRI Project MSD-2006, JPL Contract 953435 under NAS-7-100, Stanford Research Institute (1973).

[8] R. A. Howard, J. E. Matheson and D. W. North, Science **176**, 1191 (1972).

Life Sciences and Space Research XIII — Akademie-Verlag, Berlin 1975

CONSIDERATION OF PROBABILITY OF BACTERIAL GROWTH FOR JOVIAN PLANETS AND THEIR SATELLITES

D. M. Taylor, R. M. Berkman and N. Divine

Jet Propulsion Laboratory, California Institute of Technology, Pasadena, Calif., USA

Environmental parameters affecting growth of bacteria (e.g. moisture, temperature, pH, and chemical composition) were compared with current atmospheric models for Jupiter and Saturn, and with the available physical data for their satellites. Different zones of relative probability of growth were identified for Jupiter and Saturn, with the highest in pressure regions of 10^6 to 10^7 N m^{-2} (10 to 100 atmospheres) and 3×10^6 to 3×10^7 N m^{-2} (30 to 300 atmospheres), respectively.

Of the more than two dozen satellites, only the largest (Io, Europa, Ganymede, Callisto, and Titan) were found to be interesting biologically. Titan's atmosphere may produce a substantial greenhouse effect providing increased surface temperatures. Models predicting a dense atmosphere are compatible with microbial growth for a range of pressures at Titan's surface. For Titan's surface the probability of growth would be enhanced if (i) the surface is entirely or partially liquid (water), (ii) volcanism (in an ice-water-steam system) is present, or (iii) access to internal heat sources is significant.

1. Introduction

An assessment of the probability that organisms can grow in extraterrestrial environments will be described here. The approach was to categorize the factors which influence growth and relate these to the environments in question. The study has been restricted to the atmospheres of Jupiter, Saturn, and their satellites, emphasizing zones where relative probability of growth P_g, rather than mere survival, is greatest. The approach to establishing P_g values was to assess the minimal needs of bacteria and to determine whether or not atmospheric regions exist which meet these needs. Each requirement was evaluated independently in terms of its growth supporting function. Also considered was the possible existence in the atmospheres of factors which could prevent growth even in the presence of the basic growth requirements. An overall assessment in which all such factors were combined led to the establishment of zones of relatively high and low P_g.

A review of how temperature, pressure, chemical composition, and moisture affect the growth of bacteria is presented. While the maximum and minimum limits of various parameters permitting growth are described, there is no attempt to assign P_g values on the basis of the limited data. The present report derives relative P_g values as high, low, or nil. Numerical P_g values cannot be assigned reliably because large uncertainties exist in current knowledge of temperatures, pressures, abundances, turbulence, and surface characteristics for the atmospheres of Jupiter, Saturn, and their satellites.

The present P_g study used "the law of limiting factors" as a basis to calculate high, low, or nil probabilities of growth. The law states that any one factor, if limiting for growth, will affect P_g even if all other required factors are optimal.

2. Technical Discussion

2.1. Minimal Requirements for Bacterial Growth

The organisms considered in the study were bacteria rather than other life forms, primarily because of their greater variety, higher adaptability, comparative ease of study in the laboratory, and simple growth requirements in many cases, as well as the availability of extensive published literature on their physiological and morphological characteristics. A concise review of the nutritional requirements of bacteria is given by Guirard and Snell [1]. For P_g purposes, the needs have been divided into four categories: carbon source, energy source, salts, and water.

While carbon is essential for growth, bacteria can utilize a wide variety of carbon compounds, including CO_2, CH_4, and all other naturally occurring carbon compounds. Heterotrophic bacteria use the same compound as a source of energy as that which serves as the carbon source, whereas autotrophs derive energy from sunlight or by oxidation of inorganic compounds.

Bacteria require a large number of inorganic salts, either in relatively large quantity (SO_4^{2-}, NH_4^+, PO_4^{3-}, etc.), or in trace amounts (Ni^{2+}, Co^{2+}, etc.). Many salts are required both as structural entities and as co-factors in enzyme systems. While water is an absolute requirement, the growth of some micro-organisms can occur below 80% relative humidity [2], and in one case growth has been measured down to 11% relative humidity [3].

Even with all minimal growth requirements present, other inhibitory factors including radiation, temperature, pressure, and chemical agents may still limit or prevent growth. Radiation includes both ultraviolet light and energetic charged particles. In the case of temperature, the literature indicates that the growth of earth organisms cannot occur below about $-20\,°C$; the upper temperature extreme is generally considered to be near $100\,°C$ with few exceptions. The effect of pressure on bacterial growth has not been studied in detail; however, growth of marine bacteria has been shown to occur at depths corresponding to 10^8 N m^{-2} (1000 atmospheres) hydrostatic pressure [4]. Organisms have also been shown to grow under a pressure of 10^3 N m^{-2} (0.01 atmosphere) [5].

Chemical inhibitors of growth include those of a highly toxic nature as well as those which are toxic only when present in great abundance. The former includes various metabolic inhibitors such as azide and cyanide; the latter, compounds which are competitive inhibitors and those which cause secondary effects. An example of the latter is ammonia, which, if abundant, can raise the pH of aqueous suspensions to toxic levels. But some bacteria can grow at pH values approaching pH 13 [6, 7]; others are reported to grow under the acid conditions of pH 4 or lower [8].

Growth of bacteria can occur in a variety of atmospheres, including pure oxygen [9] and pure CO_2 [5, 10], in 50% NH_3 [11] and in various mixtures of NH_3, CH_4 and H_2 [11, 12].

2.2. Relevant Characteristics of Jupiter and Saturn

Available data indicate that Jupiter and Saturn (and probably other major planets) lack solid surfaces. Instead, they appear to be composed of gases whose densities decrease as a function of distance from their centers. Due to an internal heat source, large temperature gradients are predicted in their atmospheres.

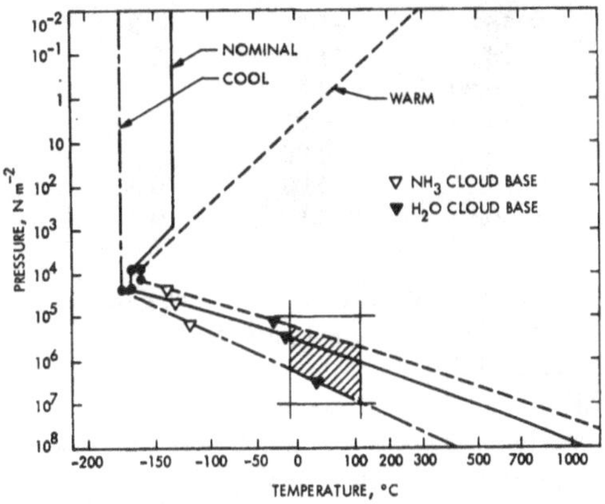

Fig. 1. Pressure vs temperature for the Jupiter model atmospheres.

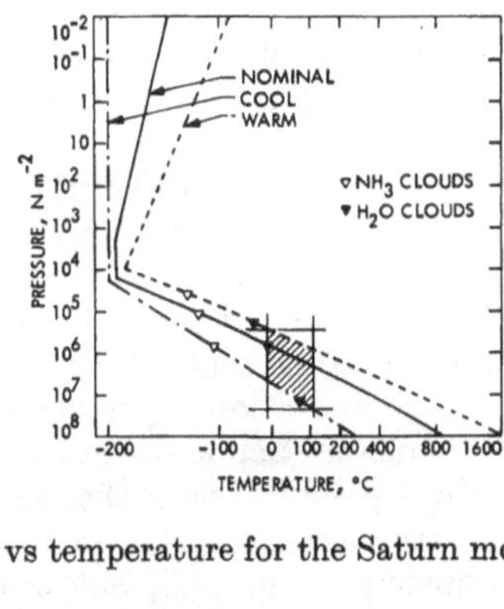

Fig. 2. Pressure vs temperature for the Saturn model atmospheres.

Although there are uncertainties in pressure-temperature relationships, models have been developed [13, 14] ranging from cool to nominal to warm (Figs. 1, 2). The models also predict the locations of aqueous and ammonia clouds. The predicted abundances of the major chemical constituents are given in Table 1.

2.2.1. *Probability of Growth in the Jupiter Atmosphere*

The cross-hatched area shown in Fig. 1 indicates the temperature range compatible with bacterial growth. It also defines a pressure range from about 10^5 to 10^7 N m^{-2} (1 to 100 atmospheres). A pressure of 10^5 N m^{-2} (1 atmosphere)

Table 1
Chemical Compositions for Jupiter and Saturn

Constituent	Jupiter, % by weight			Saturn, % by weight		
	Cool model	Nominal model	Warm model	Cool model	Nominal model	Warm model
H_2	50.7	75.3	87.7	55.3	78.5	89.5
He	46.0	23.0	11.5	39.5	19.7	9.9
H_2O	1.60	0.80	0.4	2.5	0.83	0.28
CH_4	0.86	0.43	0.21	1.33	0.44	0.15
NH_3	0.22	0.11	0.06	0.34	0.11	0.04
Ne	0.23	0.11	0.06	0.36	0.12	0.04
Others	0.4	0.2	0.1	0.7	0.24	0.08

corresponds to an equatorial distance of approximately 71400 km from the center of the planet. Significantly, the band in which temperature and pressure are compatible with bacterial growth is comparatively narrow, less than 300 km in depth. Considerations of P_g will be limited to this region. While the cool, nominal, and warm models all predict aqueous clouds in the Jupiter atmosphere, only the cool model predicts the presence of liquid water. In the cool model, liquid water should be present within the band of 2×10^6 to 5×10^6 N m^{-2} (20 to 50 atmospheres) pressure. In considering water as a limiting factor, a high P_g can probably be assigned to the 2×10^6 to 5×10^6 N m^{-2} (20 to 50 atmospheres) pressure region, and a low P_g to the remaining region between 10^5 and 10^7 N m^{-2} (1 and 100 atmospheres). However, in all three models, localized regions of high moisture content in the form of precipitation or condensation could exist even in low P_g regions.

In the region from 10^6 to 10^7 N m^{-2} (10 to 100 atmospheres), carbon is available as methane (CH_4). Ammonia is also present, although in regions below 10^6 N m^{-2} (10 atmospheres) pressure, the concentrations of methane and ammonia drop to below 0.7 and 0.2 mole m^{-3} respectively. Based on the availability of these two compounds, one can probably assign a high P_g to the 10^6 to 10^7 N m^{-2} (10 to 100 atmosphere) pressure region, and a low P_g to regions approaching 10^5 N m^{-2} (1 atmosphere) pressure. In aqueous ammonia environments, pH can play an important growth-limiting role. A pH analysis of the growth regions was conducted, based upon abundances of water and ammonia in relation to temperature. The calculated nominal pH is approximately 12.5, but would drop to about 10.5 in regions where temperatures are approaching 100°C. However, these values are based on average abundances and do not consider the effects of rains, turbulence, or other unknown factors. The other chemical constituents required for growth include several inorganic compounds such as sulfur, various divalent cations, and phosphates. Although the existence of such compounds on Jupiter is not known, it is assumed they do exist along with other inorganic compounds required in trace amounts. Radiation, chiefly in the ultraviolet wavelengths, does not pose a problem to micro-organisms since most of it is dissipated before reaching the growth region. Other kinds of radiation (X-rays, cosmic rays, etc.) are not believed to be significant in considering P_g on Jupiter. Various chemical agents inhibitory to the growth of micro-organisms may or may not be present in the Jupiter atmosphere. It is assumed that such agents will not prevent growth of micro-organisms. The assumption is based both on the attributes of bacteria and on the assumption that inhibitory agents, if present, are present only in low concentration.

Based on the above discussion, a high P_g zone is indicated in the cross-hatched band around 3×10^6 N m^{-2} (30 atmospheres). If water were not a limiting factor, the high P_g zone could have been widened to encompass a pressure zone from 10^6 to more than 5×10^6 N m^{-2} (10 to more than 50 atmospheres). Outside the 10^5 to 10^7 N m^{-2} (1 to 100 atmospheres) range, P_g is nil.

2.2.2. *Probability of Growth in the Saturn Atmosphere*

Saturn and Jupiter are markedly similar in most attributes which influence probability of growth. This is evident from Fig. 2 and from Table 1. However, there are some significant differences which will be addressed here. Because of the greater distance of Saturn, data obtained have greater uncertainties than those for Jupiter. Owing to its greater distance from the Sun, temperatures for Saturn are lower than those for Jupiter. Nevertheless, the internally generated heat source provides a region, as with Jupiter, in which temperatures are suitable for microbial growth. Pressures in this region are higher than pressures found in the analogous temperature zone identified for Jupiter. The higher pressures imply greater abundances of all chemical constituents. Fig. 2 indicates the temperature and pressure relationships for the Saturn atmosphere based on the three models. The cross-hatched area indicates where growth could be possible. The two horizontal lines confine the pressure regions to a range of about 3×10^5 to 2×10^7 N m^{-2} (3 to 200 atmospheres). Based on atmospheric analyses, liquid water should be present in the nominal model as well as in the cool model. The cool and warm models of both Jupiter and Saturn represent the extreme cases predicted for these environments. Since, for Jupiter, liquid water is predicted only for the cool model, a higher probability of growth should be assigned to Saturn.

2.2.3. *Probability of Growth on Satellites*

The satellites of Jupiter and Saturn are referred to as the "icy satellites" because of their chemical composition. While empirical measurements are relatively few, Lewis [15] has calculated the chemical composition of the icy satellites on the basis of solar chemical abundances, condensation properties of the elements, and temperatures at the time of accretion. The average calculated chemical compositions are estimated to be about 54% H_2O, 15% FeO, 13% SiO_2, 10% NH_3, 8% MgO, and less than 5% CH_4. Other compounds required for growth can be presumed present at least in trace amounts. Surface temperatures of the satellites are generally very low, probably averaging around −150 °C. The four largest Jupiter satellites, Io, Europa, Ganymede, and Callisto, are probably too cold to allow growth of micro-organisms. There is some possibility, however, that growth could occur on localized regions of the Galilean satellites if the surfaces were differentiated. Although an unlikely possibility, some surface regions could have cracks and crevices from which rich materials from the warm interiors erupt in a manner somewhat analogous to volcanic or geyser activity. This possibility is eliminated in the case of the smaller satellites which have cold interiors. The larger satellites of Jupiter have sparse atmospheres, precluding the possibility of "greenhouse effects" in which surface heat is retained. But Titan, the largest satellite of Saturn, appears to have a dense atmosphere. Thus surface temperatures are now believed to be much higher than originally thought. However, the atmosphere prevents measurement of actual surface temperatures.

At a workshop on the atmosphore of Titan, held in July 1973 [16] concensus was obtained on several physical parameters, namely a radius and mass con-

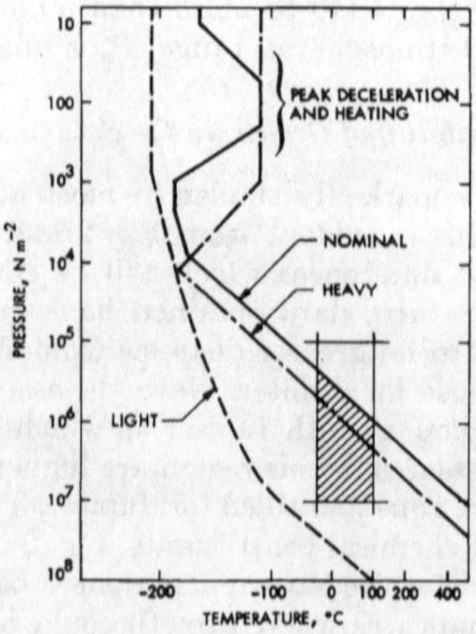

Fig. 3. Pressure-temperature profiles of engineering models of Titan's atmosphere.

siderably larger than the moon's, a gravity comparable with the moon's, and a very cold effective temperature (at 10 AU from the sun). Intensive consideration was devoted to analysis of the spectroscopic and infrared observations, and to theoretical derivations concerning the atmosphere. The only gas molecule identifiable in the spectra is methane, but there may be other major constituents, namely hydrogen, nitrogen, or argon. Water and ammonia may occur in the warmer regions of the atmosphere inaccessible to direct observations with present techniques. The data also place various constraints on atmospheric variables: for example, that surface pressures and temperatures exceed 2×10^3 N m^{-2} and -200°C respectively.

The profiles of temperature (abscissa) and pressure (ordinate) corresponding to height above Titan's surface shown in Fig. 3 include the points and theory from the workshop. The solid line represents the nominal atmospheric model and corresponds to a pure methane atmosphere (mean molecular weight $\mu = 16$ gram/mole). The dashed lines represent extreme atmosphere profiles ("heavy" composition, with $\mu = 26$, and "light" composition, $\mu = 6$). In the lower atmosphere (for pressures exceeding 10^5 N m^{-2}) a shaded region is shown. This region represents the intersection of the Titan atmospheric profiles with approximate limits on pressure and temperature for survival and growth of terrestrial organisms. This region, a few hundred kilometers thick, is the one within which a high probability of growth could occur. Note that no surface (that is, an upper bound on the pressure) is shown for this set of Titan atmosphere profiles. This occurs because present data do not establish surface pressure, but are consistent with either a thin atmosphere or a thick atmosphere; that is, surface pressures equal to or greater than 2×10^3 N m^{-2}. The profiles shown here could thus be terminated by a surface anywhere below the level at 2×10^3 N m^{-2} pressure. The growth region lies completely within those portions of the atmospheric profiles inaccessible to observation with present astronomical techniques. In this region several concerns affect growth. One of these is atmospheric dynamics (winds, weather, and other temporal and local variations), expected to be subdued on Titan because of its slow rotation (period = 16 days). Another is the likelihood of condensates. The growth region shown earlier includes only liquid ammonia as a candidate for cloud formation. However, if a surface is present, condensation of vaporized surface material, whatever it is, is likely just above the surface.

If the atmosphere profile is terminated by a surface at, for example, a pressure of 10^5 N m^{-2} and a temperature of $-120\,°C$, the surface is expected to be solid water ice with one or two major hydrates. This is the most probable surface material, and is compatible with current understanding of the structure of Titan's interior. If the surface is a solid ice crust, thermal and tidal stresses within the underlying liquid water mantle may produce the analogs of geological activity, such as volcanism, seepages, and icequakes involving water, steam, ammonia, etc. In the case of a thicker atmosphere, liquid water is the likely surface material in the growth region. This water would have ammonia in solution, at concentrations which lead to pH values between about 11 and 12.5. Such a surface ocean might serve as a growth medium for some organisms, particularly considering the availability of carbon from atmospheric methane, along with other needed materials in solution.

The large uncertainties are apparent in these atmospheric profiles and in the surface pressure. Survival and growth of some terrestrial organisms are possible if the surface pressure falls in the range of 2×10^5 to 10^7 N m^{-2} (2 to 100 atmospheres); in this case the probability of growth on Titan exceeds that for every other outer planet satellite, according to present considerations.

3. Summary

Absolute probability of growth values cannot be calculated reliably for the planetary bodies discussed here. On the basis of current knowledge, a relative rank ranging from high to nil probability of growth can be assigned to each body. Saturn has been assigned the highest probability of growth in the likelihood that liquid water and relatively high concentrations of nutrients are available. Jupiter has a somewhat lower probability of growth because there is only a 10 to 20% chance that water is present as a liquid. Titan has been ranked high on the basis of its atmosphere and the accompanying greenhouse effect. The actual surface temperature of Titan cannot be ascertained at this time because radiation measurements provide data on the upper atmosphere only. Hence, the ranking for Titan is tentative. The four Galilean satellites of Jupiter have been ranked considerably lower than Titan, and can be considered to have a very low probability of growth. Smaller satellites including the rings of Saturn, which are probably composed of ice or rock up to a few kilometers in diameter, have essentially a nil probability of growth.

Acknowledgments

This paper presents the results of one phase of research carried out at the Jet Propulsion Laboratory, California Institute of Technology, under Contract No. NAS 7-100, sponsored by the National Aeronautics and Space Administration.

References

[1] B. M. Guirard and E. E. Snell, in: The Bacteria, V. 4, Academic Press, New York 1962 (p. 33).
[2] D. J. Kushner, Adv. Appl. Microbiol. **10**, 73 (1968).
[3] S. M. Siegal and K. Roberts, Proc. Nat. Acad. Sci. **56**, 1505 (1966).
[4] C. E. ZoBell and R. Y. Morita, J. Bacteriol. **73**, 563 (1957).
[5] E. J. Hawrylewicz et al., in: Life Sciences and Space Research VI, 146 (1968).
[6] C. S. Meek and C. B. Liman, J. Gen. Physiol. **5**, 195 (1922).
[7] A. W. Downie and J. Cruickshank, Brit. J. Exptl. Path. **9**, 171 (1928).
[8] M. E. Chislett and D. J. Kushner, J. Gen. Microbiol. **24**, 187 (1961).
[9] J. Caldwell, Nature **206**, 321 (1965).
[10] J. Seckbach, F. A. Baker and P. M. Shugarman, Nature **227**, 744 (1970).
[11] S. M. Siegal and C. Giumarro, Proc. Nat. Acad. Sci. **55**, 349 (1966).
[12] S. M. Siegal and C. Giumarro, Icarus **4**, 37 (1965).
[13] The Planet Jupiter (1970), NASA SP-8069, 1971.
[14] The Planet Saturn (1970), NASA SP-8091, 1972.
[15] J. S. Lewis, Icarus **15**, 174 (1971).
[16] The Atmosphere of Titan, NASA SP-340, 1974.

Radiation Biology

Life Sciences and Space Research XIII — Akademie-Verlag, Berlin 1975

FLUX OF HIGH-LET COSMIC-RAY PARTICLES IN MANNED SPACE FLIGHT

E. V. Benton[a], R. P. Henke[a], D. D. Peterson[a], J. V. Bailey[b] and C. A. Tobias[c]

[a]Physics Department, University of San Francisco, San Francisco, Calif., USA
[b]NASA-Johnson Spacecraft Center, Houston, Texas, USA
[c]Donner Laboratory, University of California, Berkeley, Calif., USA

On the Apollo and Skylab missions the high energy heavy ion (HZE) flux was measured by means of plastic nuclear track detectors. Measurements involved the fluxes of high linear energy transfer (LET), $6 \lesssim Z \lesssim 26$ particles incident on astronauts and on several biological experiments. Partial results of these measurements are presented; the effects of shielding and solar modulation are discussed.

1. Introduction

During space missions, the radiation environment of the astronauts is quite complex. The primary galactic radiation consists primarily of protons and ^{4}He nuclei with less than 2% of the particles being high-Z, heavy nuclei. The heavy, high-Z nuclei consist primarily of nuclei of carbon ($Z = 6$) through iron ($Z = 26$) [1]. However, recently the presence of small fluxes of super heavy nuclei such as lead ($Z = 82$) and uranium ($Z = 92$) has been experimentally established [2]. The intensity and the energy spectra of these particles varies with time and position within the spacecraft. The long-term time variations come about due to the changes in the 11 year solar cycle, the rapid short term variations are due to geomagnetic effects and solar emissions, such as the solar flares. The radiation environment inside the spacecraft is also significantly influenced by the shielding offered by the various components of the spacecraft and indeed, by the bodies of the astronauts themselves [3]. For example, since the Apollo Command Module shielding varies from about 2 to 260 g cm^{-2} depending on the direction, the incoming radiation is highly anisotropic.

Background low LET (linear energy transfer) radiation is a part of everyday life; however, the high-LET radiation produced by the high Z particles is unique to the space environment. While the radiobiological action of this radiation is poorly understood, particularly when it comes to the possible long term, delayed effects, certain definite effects on biological systems have been observed. For example, it has been demonstrated that high-LET radiation is particularly effective in cell inactivation of certain non-regenerative cells (human kidney (T1) cells, for example) [4, 5]. A more dramatic effect of multicharged particles has been the light-flash phenomenon observed by the Apollo and Skylab astronauts [6—12]. It becomes important then to measure the exposure of individuals to these radiations so that this information will be available as part of the medical history and subject to future interpretations.

The technique for the measurement of dose due to protons is well understood; however, the dosimetry of HZE particles is much more complicated. Indeed, in this case the whole concept of dose is open to question. The energy associated with heavy ions is concentrated within a limited volume of tissue, so that direct effects along the particle track are far more pronounced than for lighter particles even when the absorbed dose, measured macroscopically, is the same. A knowledge of how such particle tracks are distributed within a given biological volume now becomes a significant dosimetric parameter.

2. Methods and Procedures

On Apollo and Skylab missions, the individual astronauts' HZE particle exposure was measured by means of plastic nuclear track detectors [3, 13, 14]. On Apollo, the detectors were located in the passive dosimetry packs carried on the chest, thigh, and ankle of each astronaut. On Skylab, each astronaut wore a single passive dosimeter on either the wrist or the ankle. On Apollo missions 8–12, each plastic packet consisted of two, 190 μm-thick layers of type 8070-112 Lexan. On Apollo 13 and subsequent missions, three layers were used. The addition of the third layer was found to significantly improve the charge resolution of the detector stack. Each detector has an area of about 8 cm². The layers are held stationary with respect to each others by means of a heat seal on each edge. In order to reduce the number of variables, all Lexan detectors flown to date have been taken from the same batch of Lexan. On the Skylab, in addition to Lexan, several layers of cellulose nitrate (CN-USF 4b) were added. This allowed the measurement of lower LET particles.

The track recording characteristics of CN and Lexan detectors are summarized in Table 1. Here E_{max} represents the maximum energy (in MeV/nucleon) of

Table 1

E_{max} and R_{reg} for various Ions in CN and Lexan

Particle Type	E_{max}[a] (MeV/nucl)		R_{reg} (cm)	
	CN	Lexan	CN	Lexan
^{4}He	1.5	0.2	0.004	0.0003
^{14}N	30	7	0.19	0.02
^{24}Mg	110	25	1.36	0.10
^{40}Ca	650	80	14.8	0.47
^{56}Fe	> 1000	150	> 20	1.18

[a] Corresponds to $LET_{350}^{(crit.)}$ in CN and Lexan of ~75 keV μm⁻¹ · CN and ~230 keV/μm⁻¹ · Lex respectively.

representative particles capable of registering a track in the given detector. The quantity R_{reg} represents the corresponding range of particles [15]. All detectors were processed in NaOH solution to produce conical particle tracks. Tracks are located by scanning with an optical microscope usually at 200× magnification. Each detector is scanned independently by at least two different observers. For

LET determination, measurements are performed at 600× in each of the surfaces where a visible etch cone is found. Particle identification is accomplished either by measuring the particle's LET at a given residual range (for particles stopping in the detector), or by measuring the LET and its rate of change with the residual range. It is usually possible to determine the particle's charge, energy, and its direction of travel.

For each particle which does not stop in the Lexan packet (the typical situation) the information available about this particle is the track etch rate as a function of position along the trajectory. Calibrating the detector packet consists of relating the etch rate to some function of particle atomic number Z, atomic weight A, and residual range R. Here, rather than the energy, the residual range is used as an independent variable because it is more closely related to the measurement situation. Theoretical reasoning indicates that the function of interest is approximated by the particle linear energy transfer, LET_∞. The response of plastic nuclear track detectors more closely follows the LET_{350} rather than LET_∞ of particles. This is why the quantity LET_{350} is used throughout this work. Over a wide range of particle velocities LET_{350} can be converted to LET_∞ by multiplying by a factor of ~1.6. It is found that LET_{350} represents the experimental situation well.

3. Apollo Measurements

From the HZE particle dosimetry point of view the most heavily instrumented Apollo mission was that of Apollo 17. In addition to the personnel passive dosimeters, four other biologically related experiments were instrumented to contain plastic nuclear track detectors. The four experiments included the HZE dosimeter, the Biostack, the Alfmed, and the Biocore. In addition to the Lexan, the dosimetry for these experiments also included several layers of cellulose nitrate. The detailed results from the Lexan detectors contained in the passive personnel dosimeters of Apollo 17 are given in [16], the results from the CN detectors are shown in Table 2. The approximate average shielding of each experiment is listed underneath the experiment name in column one. In columns two, three, and four appears information on the detector layer number and the measured experimental data. In columns five and six are shown the absolute planar particle track fluences. In the last column are listed the absolute planar particle flux for particles with $LET_{350}^{CN} \geqq 80$ keV μm^{-1} which corresponds to an $LET_\infty^{Tissue} \geqq 105$ keV μm^{-1}. The particle flux is in units of tracks cm^{-2} day^{-1} where the time is taken as the effective time for the mission. This is defined as the trans-earth plus trans-lunar time plus one-half of the total time spent in lunar orbit.

The data show clearly the influence of shielding on the HZE exposure with the lightly shielded HZE dosimeter recording nearly four times the flux as compared with the Biocore detectors. At the same time it is clear that even for a very heavily shielded situation, such as the Biocore (30—40 g cm^{-2} is a conservative estimate) still there are a significant number of HZE particle hits. Consequently, complete shielding from the galactic cosmic rays does not appear practical due to spacecraft weight limitations.

The data obtained from the Lexan detectors contained in the Apollo passive personnel dosimeters are shown in Figs. 1 and 2 [17]; all data represent the average for each mission. In Fig. 1 is shown the particle track fluence for particles with

Table 2
HZE Particle Track Fluences on Apollo 17[a]

Experiment	Detector layer number	Measured		Absolute planar track fluence (tr cm^{-2})		Flux[d] (tr cm^{-2} day^{-1})
		2 cones	Etched-through holes	$LET_{350}^{CN} > 170$ keV μm^{-1}[b]	$LET_{350}^{CN} > 80.0$ keV μm^{-1}[c]	$LET_{350}^{CN} > 80$ keV μm^{-1}
HZE dosimeter (2 g cm^{-2})	C5	108.5	30.8	26.1 ± 3.9	153 ± 23	16.4
Biostack (10 g cm^{-2})	E2-1	60.7	19.4	16.2 ± 2.3	95 ± 14	10.2
	E13-25	55.0	18,0	13.4 ± 1.8	78 ± 11	8.4
Alfmed (~20 g cm^{-2})	bottom sheet	26.6	14.7	11.7 ± 1.5	69 ± 9	7.4
Biocore[e] (30—40 g cm^{-2})	implanted mouse brain dosimeters	31.0	8.7	7.0 ± 1.6	41 ± 9	4.4

[a] All measurements are from cellulose nitrate of the type USF 4b processed for 10.0 hr at 40°C in 6.25N NaOH.
[b] Corresponds to an $LET_{350}^{Lexan} > 150$ keV μm^{-1} in Lexan or $LET_{\infty}^{Tissue} > 225$ keV μm^{-1}.
[c] Corresponds to an $LET_{\infty}^{Tissue} \geqq 105$ keV μm^{-1}.
[d] Flux is in effective days in interplanetary space; mission duration 301.5 hr or 9.33 effective days.
[e] The Biocore numbers are average for the four live mice.

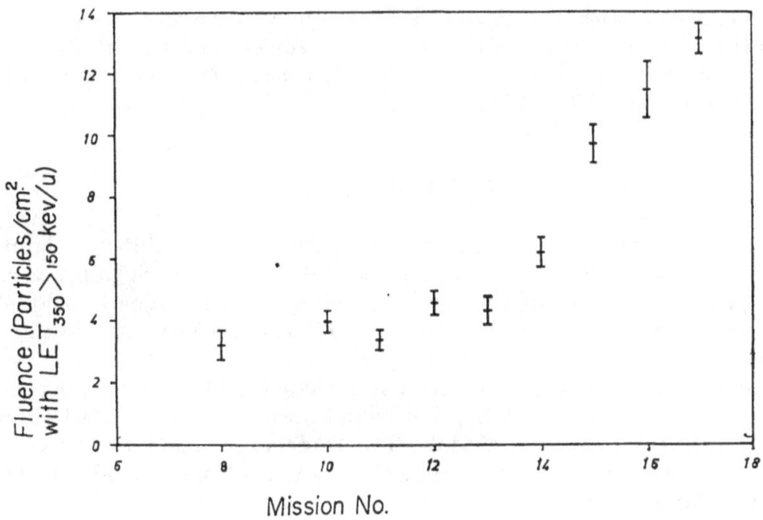

Fig. 1. The average measured particle fluence as a function of Apollo mission number.

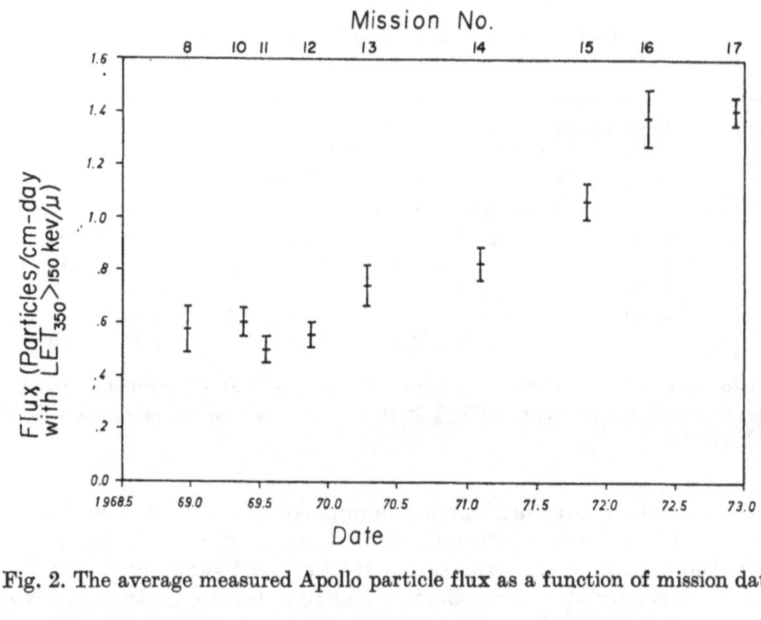

Fig. 2. The average measured Apollo particle flux as a function of mission date.

$LET_{350}^{Lexan} > 150$ keV μm^{-1} as a function of mission number. A significant increase in the fluence was observed starting with Apollo 14. Some of the increase is due simply to the increase in mission time. The absolute planar particle track flux as a function of mission date for the nine lunar missions is shown in Fig. 2. Here the time is the effective mission time as defined previously. The flux as well as the fluence is observed to increase with the Apollo 17 flux being a factor of $\sim$2.4

higher compared with the Apollo 8. This increase correlates well with solar modulation [17]. Unfortunately, Apollo 17 occurred well before the next solar minimum (during which the cosmic-ray flux becomes maximum) which will occur some time in 1975—1976.

4. Skylab Measurements

The HZE particle environment of the Skylab missions differed significantly from that of the Apollo. Being an earth orbital mission, the Skylab was shielded by the physical presence of the earth as well as the geomagnetic field. Both of these factors considerably attenuate the HZE particle flux. In addition to the personnel HZE dosimetry, which was performed on all three Skylab missions, a set of plastic nuclear track detectors was included in the D-008 experiment [8]. This experiment consisted of five individual dosimeters distributed throughout the volume of the Command Module. The D-008 experiment was flown only on the SL2 mission with the dosimeters remaining in the Command Module throughout the 28-day mission.

The measured Skylab D-008 HZE particle planar fluence and flux are listed in Table 3. Here, the first column designates the dosimeter, the second, the approxi-

Table 3
HZE Particle Planar Fluence and Planar Flux
in the Skylab (D-008-SL2) Command Module[a]

Dosimeter (layer 11)	Approximate shieldng (g cm^{-2} of Al)	Fluence (particles cm^{-2})		Planar flux[b] (particles cm^{-2} day^{-1})
		Observed	Planar[b]	
F1	~ 10	58.2 ± 7.6	93.8 ± 9.7	3.33 ± 0.38
F2	> 20	35.1 ± 5.9	27.5 ± 5.2	1.01 ± 0.19
F3	> 20	26.1 ± 5.1	21.1 ± 4.6	0.75 ± 0.19
F4	~ 10	53.4 ± 7.3	85.9 ± 9.3	3.08 ± 0.31
F5	~ 20	52.8 ± 7.3	37.0 ± 6.1	1.32 ± 0.19

[a] The data represent measurements performed with type USF 4b cellulose nitrate.

[b] Corresponds to particles with $LET_{350}^{CN} \geqq 80$ keV μm^{-1} which is equivalent to $LET_{\infty}^{Tissue} \geqq 105$ keV μm^{-1}.

mate shielding. The third and fourth columns contain the particle fluence data with the planar fluence being corrected for detector efficiency. The planar flux is shown in the last column and represents the fluence divided by the total mission time. The data clearly show the effect of shielding on the particle flux with the flux changing from a high of 3.33 to a low of 0.75 particles cm^{-2} day^{-1}. The average planar flux for the five dosimeters for particles with $LET_{350}^{CN} \geqq 80$ keV μm^{-1} ($LET_{\infty}^{Tissue} \geqq 105$ keV μm^{-1}) is 1.90 ± 0.30 particles cm^{-2} day^{-1}.

In Table 4 is shown the comparison between the Apollo 17 and Skylab (D-008) measurements. Since the Skylab utilized a slightly modified version of an Apollo Command Module, the comparison between the two is meaningful. The Apollo flux of particles with $LET_{350}^{CN} \geqq 80$ keV μm^{-1} is nearly 5 times higher than that

Table 4

Comparison of the Average HZE Particle Planar Flux in Skylab (SL2) and Apollo 17

Mission	Average HZE particle planar flux density (cm^{-2} day^{-1})	
	$LET_{350}^{CN} \geqq 170$ keV μm^{-1} [a]	$LET_{350}^{CN} \geqq 80$ keV μm^{-1} [b]
Skylab (SL2) (Command Module)	0.13 ± 0.04	1.90 ± 0.30
Apollo 17	1.58 ± 0.22	9.27 ± 1.33

[a] Corresponds to $LET_{350}^{Lex} \gtrapprox 150$ keV μm^{-1}; $LET_{\infty}^{Tissue} \gtrapprox 225$ keV μm^{-1}.
[b] Corresponds to $LET_{350}^{Lex} \gtrapprox 70$ keV μm^{-1}; $LET_{\infty}^{Tissue} \gtrapprox 105$ keV μm^{-1}.

of Skylab. A similar ratio for particles with $LET_{350}^{CN} > 170$ keV μm^{-1} is about 12. As stated previously the differences arise from earth shielding and the geomagnetic effect; however, there may also be a solar modulation factor. The differences in the ratios for the different LET's also imply differences in the integral LET particle spectra for the two kinds of missions. The integral LET particle spectrum is definitely steeper in the case of the Skylab. This reflects the lower incidence of the highest LET particles which are filtered out, because of their generally lower energy, by the geomagnetic field.

5. Summary and Conclusions

The most interesting results and observations obtained to date with plastic nuclear track detectors can be summarized as follows:

1. Only the fraction of the $Z \geqq 6$ particle flux that corresponds to essentially stopping particles was measured. These are the highest LET particles, $LET_{\infty}^{Tissue} \gtrsim 100$ keV μm^{-1} (and probably the most biologically significant), that either stop in the detector or the astronaut.
2. Comparison of track fluences measured on later Apollo missions with those from earlier missions indicates an increase in the HZE cosmic-ray flux. This increase is partially due to a decrease in the degree of solar modulation.
3. The Apollo flux of high-LET, HZE particles is considerably higher than that of Skylab.
4. Considerable variations have been found to exist in the recorded high-Z track fluences as a function of detector location on the astronaut's body and the spacecraft implying that shielding is important.
5. Although the plastic track detectors are heavily biased toward the higher Z particles, fewer $Z = 26$ (Fe) particles are observed than expected as compared with lighter particles. This suggests a break-up of Fe particles in passage through the spacecraft shielding. Thus the Z spectrum inside the spacecraft is shifted toward the lighter particles.
6. From these measurements, stopping particle densities in the form of integral Z distributions can be obtained and integral LET spectra can be measured and used in estimating biologically meaningful quantities.

7. The frequency of the light flash phenomenon reported by the astronauts is considerably higher than the frequency with which the $LET_{\infty}^{Tissue} \geqq 100$ keV μm⁻¹ particles are incident on tissue as recorded with plastic detectors. This implies that lower LET particles are at least partially responsible.

References

[1] D. GRAHN, National Academy of Sciences Report (1973).
[2] P. H. FOWLER, Proc. 13th Int. Cosmic Ray Conf., Denver **5**, 3627 (1973).
[3] E. V. BENTON und R. P. HENKE, Air Force Weapons Laboratory Rep. AFWL-TR-72-5 (1972).
[4] P. TODD, Rad. Res. Suppl. **7**, 196 (1967).
[5] University of California Radiation Laboratory Rep. UCRL-17357; Lawrence Berkeley Laboratory Reps. LBL-529 (1971) and LBL-2016 (1973).
[6] C. A. TOBIAS, University of California Radiation Laboratory Rep. UCRL-8915 (1958).
[7] G. G. FAZIO, J. V. JELLEY and W. M. CHARMAN, Nature **228**, 260 (1970).
[8] C. A. TOBIAS, T. F. BUDINGER and J. T. LYMAN, Nature **230**, 596 (1971).
[9] T. F. BUDINGER, C. A. TOBIAS, J. T. LYMAN, P. K. CHAPMAN, L. S. PINSKY, H. BICHSEL, J. D. DENNEY and W. B. NELP, Lawrence Berkeley Laboratory Rep. LBL-31 (1971).
[10] T. F. BUDINGER, J. T. LYMAN and C. A. TOBIAS, Nature **239**, 209 (1972).
[11] P. J. MCNULTY, V. P. PEASE, L. S. PINSKY, V. P. BOND, W. SCHIMMERLING and K. G. VOSBURGH, Science **178**, 160 (1972).
[12] L. S. PINSKY, W. Z. OSBORNE, J. V. BAILEY, R. E. BENSON and L. F. THOMPSON, Science **183**, 957 (1974).
[13] E. V. BENTON and R. P. HENKE, University of San Francisco Reps. USF-TR-71-15 (1971), USF-TG-72-20 (1972), USF-TR-73-26 (1973).
[14] D. D. PETERSON and E. V. BENTON, University of San Francisco Rep. USF-TR-73-29 (1973).
[15] E. V. BENTON, U. S. Naval Research and Development Laboratory Rep. USNRDL-TR-68-14 (1968).
[16] E. V. BENTON and R. P. HENKE, Health Physics **27**, 79 (1974).
[17] E. V. BENTON and R. P. HENKE, University of San Francisco Rep. USF-TR-74-30 (1974).

Life Sciences and Space Research XIII — Akademie-Verlag, Berlin 1975

THE STUDY OF THE RADIATION ENVIRONMENT IN NEAR-EARTH SPACE

V. M. Petrov, Y. A. Akatov, S. B. Kozlova, V. V. Markelov, V. M. Nesterov, V. I. Redko, L. N. Smirenny, A. V. Khortsev and I. V. Chernikh

Institute of Biomedical Problems, Moscow, USSR

The report contains the description of the devices used for studying the radiation environment in space. They consist of passive dosimeters, a monitoring dosimeter and a spectrometer. Data were obtained with them over a long period of time. The analysis of these data permits one to conclude that radiation dose greatly depends on the apogee altitude and inclination; increasing the shield thickness does not greatly decrease the daily dose. The daily doses in orbits with an inclination of less than 65° are 7 to 45 mrad day^{-1}; the quality factor in those orbits was 1.2—1.4.

1. Introduction

Radiation safety measures for cosmonauts require measurement of radiation doses in space vehicles in orbit. Radiation measurement in possible orbits of such vehicles over long periods of time are of great importance, and give opportunities to understand the dependence of radiation doses on the physical conditions in space.

2. Radiation Monitoring System

The system for radiation doses measurement consists of passive dosimeters and devices for measurement of characteristics of cosmic rays along their trajectories. Passive dosimeters provide a measurement of dose per mission. The passive dosimetry package (in size 65 mm × 47 mm × 13 mm) contains thermoluminescent glass dosimeters and nuclear emulsions [1]. On the satellites these packages were located in certain points inside the modules. On the manned spaceships the packages were placed in the pockets of the cosmonauts' constantly worn garments. During the first flights of the spaceships Vostok and Voshod the passive dosimetry packages had in addition photo-dosimeters and radiophotoluminescent tablets [2].

Modern passive dosimetry packages provide measurement of the absorbed radiation dose in the range of 50 mrads to 5000 rads with an accuracy to ±15%. The nuclear emulsions provide measurements of the cosmic radiation fluxes and spectra that are necessary for determining the quality factor of the absorbed cosmic radiation.

For these purposes nuclear emulsions of different sensitivity are located in the packages. Emulsions of types H-3 and H-4 have low sensitivity and are used for the measurement of multi-charged particles and low energy protons [3]. Я-2 emulsions having medium sensitivity are used for the measurement of protons with energy up to 100 MeV. By using special methods of developing the emulsions one can obtain linear energy transfer (LET) distributions of particles with up to 1 GeV.

Having estimated the average value of the LET for all types of particles it is possible to estimate the average quality factor (QF) of the absorbed radiation using the equation

$$\overline{\mathrm{QF}} = \frac{\sum_{i=1}^{n} D_i^{\mathrm{tis}}\,\mathrm{QF}}{\sum_{i=1}^{n} D_i^{\mathrm{tis}}}.$$

In this equation the dose D_i from the particles of ith type is

$$D_i^{\mathrm{tis}} = K \left(\overline{\frac{dE}{dx}}\right)_i N_i$$

where $\overline{dE/dX}$ is the average LET of the ith-type particles. The values of QF are taken from [4].

For tissue-dose measurements monitoring dosimeters with tissue-equivalent ionization chambers are used. The filter shielding the chamber permits one to measure doses behind shielding screens with thickness of 0.3 g cm^{-2} or 3.0 g cm^{-2}. The ionization chamber measures a range of dose rate of 5×10^{-4} to 6×10^2 rad hr^{-1}. The ionization chamber walls are made of tissue-equivalent plastic, and the chamber is filled with tissue-equivalent gas mixture (64% C_2H_2, 26% CO_2, 10% N_2).

For measuring the spectra of electrons, protons, alpha particles and nuclear particles with $Z > 2$ a monitoring spectrometer is used. These particles are registered in pre-arranged differential energy bands. The width of the differential energy band is fixed in such a way that changes of the QF in each energy band are slight.

The energy ranges of particles detected by the spectrometer are: for electrons, 0.5—3.0 MeV; for protons, 7—250 MeV; for alpha particles, 28—500 MeV.

The sensor of the spectrometer consists of two solid-state detectors. The minimum energy of the particles registered is determined by the thickness of the first detector, 800 μm. In this case the pulse detector amplitude exceeds the noise preamplifier level. Particle type and energy determination is made using dE/dX and total energy deposition, or dE/dX and dE/dX techniques. The pulses from both detectors are transmitted to the matrix of the discriminators. The pulse coincidence in certain discriminators determines the type and energy of the particles. This matrix has 15 outputs, forming groups corresponding to certain particle types. The number of pulses in these groups is telemetered to the ground.

3. Results

The dose and spectral measurements were performed on the Cosmos satellites and on manned vehicles.

The major part of the data has been obtained from passive dosimeters. For more detailed investigation of the radiation environment in new orbits electronic devices have been used. The cosmic radiation doses per day obtained during 1961—1973 are given in Table 1 and Fig. 1. The orbit parameters are taken from [5]. Altitude of the apogee and the inclination are stated in round numbers. The radiation measurements on the cosmic station Salyut are from electronic devices.

The station Salyut was put into orbit in June 1971 with apogee of 280 km and inclination of 52°. During the flight the crew of the station turned on the dosimeter and the spectrometer according to the flight programme, and measured daily dose and spectra of the cosmic radiation. The doses were 11 to 14 mrad day^{-1}. Particle fluxes per day in different energy bands are given in Table 2. The varia-

Table 1
Dose Measurement Data

Date		Orbit parameters		Dose	References
month	year	Inclination	Apogee (km)	(mrad day^{-1})	
08	1960	65°	340	12.5	[6]
04	1961	65°	330	7.2	[2]
08	1961	65°	240	8.4	[2]
05	1962	65°	330	16.0	[7]
08	1962	65°	370	45.0	[7]
08	1962	65°	240	13.5	[2]
10	1962	65°	350	30.0	[7]
11	1962	65°	380	30.0	[7]
12	1962	65°	400	35.0	[7]
04	1963	65°	340	18.0	[7]
05	1963	65°	370	15.0	[7]
05	1963	65°	400	30.0	[7]
06	1963	65°	300	12.0	[2]
06	1963	65°	220	15.5	[2]
06	1963	65°	230	15.5	[2]
10	1964	65°	410	29.0	[2]
04	1965	65°	500	65.0	[2]
09	1965	65°	360	16.0	this paper
10	1965	65°	340	15.0	,,
11	1965	65°	350	18.0	,,
03	1966	65°	310	11.0	,,
04	1966	65°	330	15.5	,,
07	1966	73°	350	21.0	,,
07	1966	52°	360	26.0	,,
08	1966	65°	360	29.0	,,
11	1966	65°	340	22.0	,,
12	1966	65°	320	14.0	,,
11	1968	52°	~210	21.2	,,
01	1969	52°	~210	20.0	,,
01	1969	52°	~210	24.0	,,
10	1969	52°	~210	14.0	,,
10	1969	52°	~210	13.0	,,
10	1969	52°	~210	14.0	,,
06	1970	52°	~210	13.0	,,
12	1973	52°	~210	12.5	,,

Table 2
LET-spectra Measurements Data

LET (MeV cm^{-1})	Particle flux (part cm^{-2} day^{-1})	Daily Dose (mrad day^{-1})	(mrem day^{-1})
<5.8	6.4×10^4	4.5	4.5
11–5.8	1.6×10^4	2.1	2.1
19–11	3×10^3	0.7	0.7
25–19	6.5×10^2	0.2	0.2
32–25	1.5×10^3	0.7	0.7
43–32	3.3×10^3	1.9	1.9
219–85	3.7×10^2	0.7	2.0
Z>2	60	0.2	1.3
Total dose		11	13.4

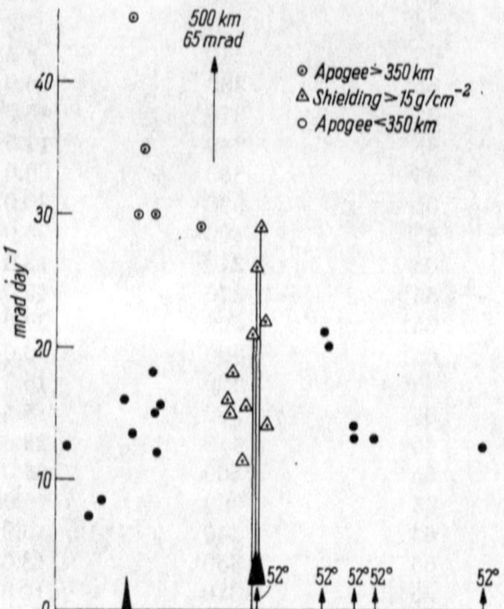

Fig. 1. Values of the average doses per day obtained on the Cosmos satellites and on manned vehicles.

tions during the flight were slight, so the values given can be considered as typical of the radiation environment in the orbits of the spaceships Soyuz and Salyut.

Linear energy transfer distributions of cosmic rays are of great interest because from these distributions it is possible to determine the QF. These distributions obtained from nuclear emulsions during the flights of the vehicle Soyuz are given in Fig. 2.

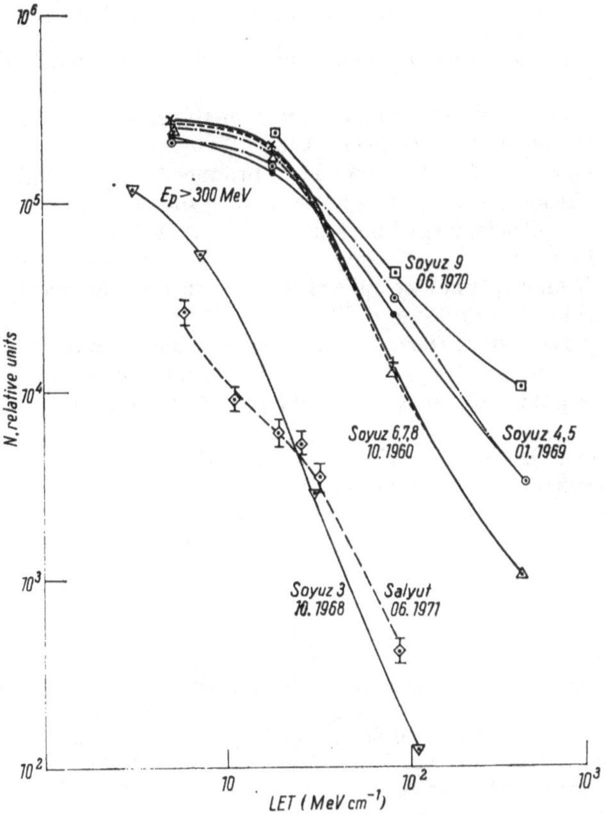

Fig. 2. The LET spectra of the particles obtained on board the manned vehicles.

4. Analysis of the Measurements

The results of the daily doses are presented in Fig. 1. The inclinations are given if they are different from 65°. A number of measurements were made to estimate the shield efficiency in near-earth orbits. The shield thickness in these experiments exceeded 15 g cm^{-2}.

The results confirmed earlier conclusions on the dependence of the dose on the period of the solar activity cycle. During the solar minimum the doses increase by a factor of about 2. Rather large values obtained in 1963 are probably connected with the radiation that was artificially injected by high altitude nuclear tests in 1962 [9]. The doses measured behind the considerable shields (in the range of 12 to 18 mrad day^{-1}) are perhaps due to high energy protons of galactic cosmic rays and the radiation belts. Anomalously large values measured in July and August 1966 can be explained by the penetration of the high energy protons of the solar flares (of 30th July and 28th August) into the vehicle. The spectral characteristics of the protons generated in those flares (the characteristic rigidities were 400 MV and 317 MV respectively [10]) were the cause of the increase even behind the shield of thickness 15 g cm^{-2}.

From these analyses one may draw the following conclusions:

(i) the dose in orbits with an inclination of 52° to 73° greatly depends on the apogee altitude;

(ii) the dependence of dose on the period in the solar activity cycle is weaker than its dependence on the apogee altitude;

(iii) the radiation within the vehicle is produced by the high energy cosmic rays passing through the geomagnetic fields and by the radiation belts, so that an increase of shield thickness from 2 g cm^{-2} to 15 g cm^{-2} does not greatly decrease the dose rate;

(iv) the high energy solar cosmic rays can cause an increase of dose rate in orbits with inclinations of 65° and 52°;

(v) the daily doses from these factors are 7 to 45 mrad day^{-1};

(vi) the major item in the dose measured in rem is given by the high energy protons of the galactic cosmic rays and radiation belt in the South Atlantic Anomaly;

(vii) the mean QF in those orbits was 1.2—1.4;

(viii) the results obtained confirm the earlier conclusion about the radiation safety of cosmic flights in orbits with inclinations of less than 62° and apogee of less than 350 km provided there are no powerful solar flares.

References

[1] Y. A. Akatov et al., Proc. Int. Congr. on Protection against Accelerator and Space Radiation 1, 492 (1971).

[2] P. P. Saksonov et al., Essays of the Cosmic Radiobiology, Nauka, Moscow 1968 (p. 494).

[3] K. S. Bogomolov et al., Kosm. Biol. i Med. 1, No. 3, 39 (1967).

[4] Recommendations of the International Commission on Radiation Protection, Atomizdat, Moscow 1958.

[5] Cosmonautics, The Small Encyclopedia, Sovyetskaya Encyclopedia, Moscow 1968 (p. 466).

[6] I. B. Keirim-Markus et al., Artificial Satellites, 12, 47 (1962).

[7] S. N. Vernov et al., Kosm. Issled. 2, 136 (1964).

[8] V. A. Blashkovsky et al., Kosm. Biol. i Med. 5, No. 2, 50 (1971).

[9] The Starfish Operation, Atomizdat, Moscow 1964.

[10] W. Atwell, Proc. Nat. Symp. on Natural and Manmade Radiation in Space, NASA IMX-2440, Washington, D.C. 1972 (p. 329).

Life Sciences and Space Research XIII — Akademie-Verlag, Berlin 1975

PHYSICAL DOSIMETRIC EVALUATIONS IN THE APOLLO 16 MICROBIAL RESPONSE EXPERIMENT

G. R. TAYLOR[a], J. V. BAILEY[a] and E. V. BENTON[b]

[a] Life Sciences Directorate, NASA-Johnson Space Center, Houston, Texas, USA
[b] Department of Physics, University of San Francisco, San Francisco, Calif., USA

Nine biological species, including viruses, bacteria, fungi, and nematodes, were exposed to various combinations of space vacuum, galactic radiation, and solar UV light during the Apollo 16 space flight. No major changes in number of surviving cells occurred, permitting detailed genetic and somatic studies of returned test subjects. To enable dose-response studies, solar UV was employed as a mutagenic source with cells exposed to full sunlight or to components of the UV spectrum at peak wavelengths of 254, 280, and 300 nanometers over a range of energy levels. Proper in-flight UV irradiation monitoring required the development of a potassium ferrioxalate actinometer and an anaerobic photographic emulsion dosimeter which were tested for the first time in space. Studies of the mutagenic activity of cosmic-ray particulate radiation environment required measurement of its components with several lithium fluoride thermoluminescent dosimeters and a package of passive nuclear-track detectors capable of recording high-energy multicharged particles. These detectors included cellulose nitrate, Lexan, Ilford G5, and silver chloride crystals. The nuclear track detectors measured the incident heavy particles with the recorded spherical fluences with $LET_{350,H_2O} > 100$ keV $\times \mu m^{-1}$ to be 19.3 ± 1.8 particles cm^{-2}. This value was found to be lower than that recorded by detectors located in the Biostack and the passive personnel dosimeters worn by the astronauts, suggesting a somewhat greater average shielding.

1. Introduction

Early studies conducted by the United States and the Soviet Union have established the ability of microbes to survive the space environment, although actual survival rates of space-flown cultures were often different from analogous ground controls. More recent space flight studies have produced data that suggest the possibility of space-flight-mediated genetic and developmental alterations in several species. The results of these earlier studies, which have recently been reviewed by Taylor [1], prompted the complex of biological studies known as the Microbial Response to Space Environment experiment. This experiment, flown on the Apollo 16 spacecraft, was designed to evaluate the relationship between space flight and viability and to study the effect of space flight on phenotypic expressions in several biological systems [2].

Table 1
Components of the Apollo 16 Microbial Response to Space Environment Experiment System

Biological Components

Phenomenon studied	Assay system	Micro-organism	Investigator
Lipolytic α toxin production	Lytic zone on agar	*Bacillus thuringiensis*	R. T. Wrenn, W. L. Ellis Northrop Services, Inc. Houston, Texas
Deforming β toxin production	*Sarcina flava* and house fly		G. R. Taylor, R. C. Simmonds NASA Johnson Space Center Houston, Texas
Fatal δ toxin production	Silk worm and crystal assay		A. M. Heimpel U. S. Dept. of Agriculture Beltsville, Maryland
Infectivity	Mouse	*Nematospiroides dubius*	R. A. Long, W. L. Ellis Northrop Services, Inc. Houston, Texas G. R. Taylor NASA Johnson Space Center Houston, Texas
Hemorrhagic factor production	Guinea pig and hemoglobin	*Aecromonas proteolytica*	B. G. Foster, D. O. Lovett Texas A. & M. University College Station, Texas
Hemolytic enzyme production	Human erythrocytes		
Genome alteration	Spore production	*Bacillus subtilis* spores, strains HA 101 and HA 101 (59) F	J. Spizizen, J. E. Isherwood Scripps Clinic and Research Foundation La Jolla, California
UV and vacuum sensitivity	Colony formation	*Bacillus subtilis* spores, strain 168	H. Bücker, G. Horneck, H. Wollenhaupt University of Frankfurt Germany
Bacteria phage infectivity	Host lysis	*Escherichia coli* (T-7 phage)	J. Spizizen, J. E. Isherwood Scripps Clinic and Research Foundation La Jolla, California
Cellulolytic activity	Cloth fibers	*Chaetomium globosum*	P. A. Volz, Y. C. Hsu, D. E. Jeger, J. L. Hiser, J. M. Veselenak Eastern Michigan University Ypsilanti, Michigan
Animal tissue invasion	Human hair	*Trichophyton terrestre*	
Drug sensitivity	Antibiotic sensitivity in agar	*Rhodotorula rubra* *Saccharomyces cerivisiae*	

Table 1 (Continuation)
Dosimetry Components

Measurement	Monitor used	Assay systems	Investigator
High-energy multicharged particles	Passive nuclear track detectors	Lexan Cellulose nitrate Photographic emulsion Silver chloride	E. V. Benton University of San Francisco San Francisco, California
Ultraviolet light	Passive dosimeters	Potassium ferrioxalate actinometry Photographic emulsion	M. B. Parson, R. A. Long, W. Ellis Northrop Services, Inc. Houston, Texas G. R. Taylor NASA Johnson Space Center Houston, Texas
Penetration of galactic irradiation	Thermoluminescent dosimeters	Lithium fluoride	J. V. Bailey NASA Johnson Space Center Houston, Texas R. A. English, R. D. Brown Kelsey-Seybold Clinic Houston, Texas

2. General Experimental Design

The biological phenomena studied in the microbial response experiment are outlined in Table 1. These represent well-known model systems that can be correlated with disease or other medically important conditions that could affect the health of future astronauts. Experts were invited to study these phenomena within their area of expertise and to conduct critical investigations in their own laboratories. Each investigator selected a species that was non-pathogenic to man, was well suited to simple screening tests, was compatible with the flight hardware environment, and was well characterized relative to the phenomenon to be studied. The first series of studies are reviewed in [3]. The results of the detailed biological studies will be presented elsewhere.

3. Physical Dosimetry Requirements

3.1. Ultraviolet Light Dosimetry

During the space flight, some of the test system replicates were exposed to UV components of the ambient solar spectrum so that the response of stressed biological systems could be compared with analogously stressed systems in the laboratory. An optical system, described in [4], was provided within the flight hardware which permitted exposure of the test systems to the full light of space or to components of the solar UV spectrum. Peak wavelengths of 254, 280 and 300 nanometers were selected by using quartz bandpass interference filters. In addition quartz neutral-density filters controlled the radiant energy reaching the exposed system. This energy ranged from a minimum of 40 ergs cm^{-2}* to a full sunlight penetration of 8×10^8 ergs cm^{-2}. The total UV flux, penetrating various combinations of filters, was monitored by the two methods outlined below.

3.1.1. Photographic Film Actinometry

Kodak High Resolution SO-343 photographic film was calibrated to measure irradiation at a peak wavelength of 254 nanometers. To minimize image loss and to increase the reliable time interval between exposure and processing, the oxygen and moisture were removed at 1×10^{-5} torr (1.33×10^{-3} N m^{-2}) for one hour and replaced with dry nitrogen gas. This process resulted in a sensitization of the film so that it was useful over a range of 3 to 120 ergs cm^{-2}. Although this range was below the level of biological interest, especially dense filters were designed so that the value of this monitoring method could be evaluated in flight. The system proved to be of value in confirming the variability of the very dense special optical filters but was too sensitive to be of practical value for the biological systems. Subsequent use of this monitoring method would require additional calibration with a different optical system.

3.1.2. Potassium Ferrioxalate Actinometry

Potassium ferrioxalate solution was selected to measure solar energy at peak wavelengths of 254, 280, and 300 nanometers because of its sensitivity and linear response to the middle ultraviolet regions [5]. The solution was prepared stoichiometrically by diluting 2 cm^3 of a 0.2 molar ferric ammonium sulfate solution and 2 cm^3 of a 0.6 molar potassium oxalate solution with 50 cm^3 of 0.1 N sulfuric acid. Calibration tests conducted with the X-25 solar simulation source at the Jet Propulsion Laboratory in Pasadena, California, demonstrated that this solution was sensitive to UV irradiation in the range between 5×10^3 ergs and 4×10^5 ergs. During exposure Fe III (ferric) ions were changed to Fe II (ferrous) which were found to be most stable at pH 1.5. Since the remaining Fe (II) decay rate was temperature dependent a rate constant was applied to the final data to allow for temperature variation. Several recording thermometers within the flight hardware verified the nearly constant temperature maintained throughout the mission. Post-flight, the various replicates of in-flight test, in-flight control, ground control, and launch vibration control aliquots (0.05 cm^3 each) were visualized by reacting with buffered phenanthroline and compared optically with suitable blanks.

The applicability of the system for in-flight UV actinometric monitoring was verified by inclusion of pre-irradiated aliquots in the flight hardware. The calculated energy level varied less than $\pm 7\%$ within the flight group and between this set and the analogous ground control groups. These data indicate two things: (i) neither the space flight environment, nor the simulated launch vibration exerted a detectable change in the pre-irradiated ferrioxalate (when compared with suitable ground controls); (ii) the reduced reaction product in the irradiated potassium ferrioxalate solution was sufficiently stable to allow for a post-flight

* 1 erg = 10^{-7} joule.

estimation of the irradiation received in flight, thus allowing verification that the device was exposed as planned.

Post-flight measurement of in-flight irradiated aliquots most often reflected values within the established range of $\pm 7\%$ of the anticipated value. In some aliquots, however, measured energies were from two to three times the anticipated quantities. These higher-than-anticipated energies were considered to have been made possible by stress lines and other anomalies in the optical system which resulted from the 100 inch-ounce (7.9×10^{-3} cm gram) force used to seal the hardware components. This variable does not invalidate any experiment but does signal the fact that some replicates may have received more light than others. Most experiments produced growth-response curves over a range of several log units with many replicates being used under several different filters. Under these conditions "hot spots" could be detected, if present, the resulting data being evaluated accordingly.

3.2. High-Energy Multicharged Particle Dosimetry

The high-energy multicharged cosmic-ray particle exposure during the Apollo 16 mission was monitored using cellulose nitrate, Lexan polycarbonate, nuclear emulsion, and silver chloride crystal* nuclear track detectors.

Both the Lexan and the cellulose nitrate (CN) detectors revealed track fluences (tracks cm^{-2}) of the HZE particles. Since the CN detector is more sensitive, it showed track fluences substanitally higher than that found in Lexan. The sensitivity of the two detectors is such that the CN records particles with a Z (atomic number) greater than 6, while Lexan records particles with a Z greater than 10. The scanned cellulose nitrate layer yielded 109 tracks due to the heavy, high-energy cosmic-ray particles in its 6.82 cm^2 area yielding a measured fluence of 16.0 ± 1.5 tracks cm^{-2}. Of these tracks, 19 registered in the immediately adjacent Lexan layer, yielding a measured fluence of 2.8 ± 0.6 tracks cm^{-2}. Track length measurements in both detectors were reduced to integral LET_{350} spectra [6].

Comparison of Lexan and CN track fluences found in the flight hardware showed them to be somewhat lower than that found in either the light flash detector or the passive personnel dosimeters [7]. The Apollo 16 passive personnel dosimeter mission average fluence of particles with $LET_{350,Lexan} > 150$ keV μm^{-1} was 19.4 ± 0.8 tracks cm^{-2}. (The response of plastic nuclear track detectors more closely follows the LET_{350} rather than LET_{∞} of particles. This is why the quantity LET_{350} is used throughout this work. Over a wide range of particle velocities LET_{350} can be converted to LET_{∞} by multiplying by a factor of ~ 1.6.) This value was four times larger than the corresponding fluence of 4.9 ± 0.5 tracks cm^{-2} in the microbial response hardware.

These observations, along with the depressed thermoluminescent dosimetry (TLD) values presented below, imply that the microbial response flight hardware had a somewhat greater average shielding compared with either the light flash detector or the personnel passive detectors. Likewise these data are slightly lower than those obtained from the TLD and CN detectors employed in the Biostack flight hardware, which was stowed in the Command Module in an area of minimal shielding to ambient cosmic radiation [8]. Within the microbial response experi-

* Silver chloride crystals courtesy of E. Schopper, University of Frankfurt, and C. Childs, University of North Carolina.

ment, the spherical fluence of particles with $LET_{350,H_2O} > 100$ keV μm^{-1} was found to be 19.3 ± 1.8 particles cm^{-2}. The planar fluence with $LET_{350,Lexan} > 150$ keV μm^{-1} was 4.9 ± 0.5 particles cm^{-2}. The arbitrary cutoff value of $LET_{350,Lexan} = 150$ keV μm^{-1} is in accordance with previous usage and is used here to allow objective fluence comparisons. Ilford G5 and silver chloride crystals were used as a backup system to be reported in case of trouble with the Lexan and cellulose nitrate. Because the operation of the prime system was completely satisfactory, the backup option was not exercised.

3.2. Thermoluminescent Dosimetry

Seventy-six lithium fluoride thermoluminescent [9] chips (5 mm × 5 mm × 1 mm) were included in the flight hardware to measure the integrated dose of ionizing radiation to the test components deriving from the broad spectrum space radiation environment. Preflight radiation calibration was performed against ^{60}Co with subsequent analyses of flight chips, control chips, and calibration chips being conducted 20 days after termination of the Apollo 16 mission.

The TLD chips were arranged within the flight hardware in packets of 4 chips each as shown in Fig. 1. The unvented tier ran the full length of the device along

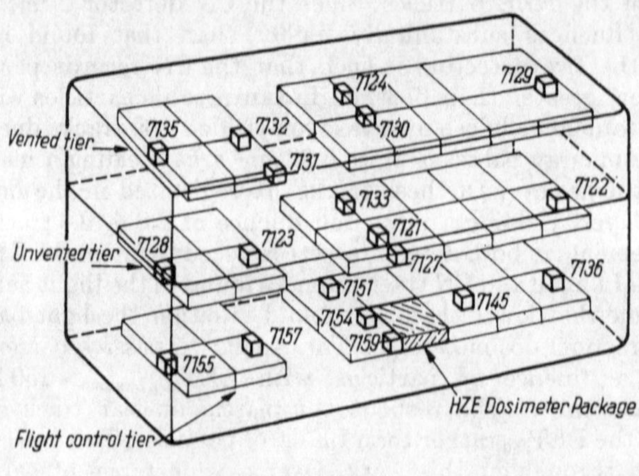

Fig. 1. Arrangement of thermoluminescent and high-energy multicharged particle dosimeters in the Apollo 16 microbial exchange hardware. Numbers correspond to TLD sample numbers in individual sample chambers (cuvettes) and represent 4 chips each.

the central axis and was therefore slightly more shielded from ionizing radiation than the outer layers. Only the flight control tier was not exposed to solar UV radiation. This distribution allowed dose determination for each tier, for each of the six sides of the flight device, and for the central volume of the closed assembly.

The ionizing radiation dose data were analyzed statistically. The flight hardware received extremely uniform irradiation from the ionizing radiation components of the space environment. The doses on the several tiers had means and standard deviations as follows: unvented tier, 0.49 ± 0.01 rad*; vented tier, 0.47 ± 0.02 rad; flight control tier, 0.48 ± 0.02 rad; the dose between different chips on one tier differed by, at the most, 0.05 rad, usually only 0.01—0.02 rad [4]. The mean ionizing radiation dose reaching the test systems was 0.48 ± 0.02 rad. By comparison, crewmen received a mean ionizing radiation dose of 0.51 ± 0.02 rad which implies a greater average shielding for the Microbial Exchange experiment. TLD materials from both sources were calibrated concurrently but were analyzed independently to allow accurate and unbiased comparison of results.

The measured dose of 0.48 ± 0.02 rad represents an absorption of 48 ± 2 ergs of ionizing energy per gram of biological material within the flight hardware. This value is applicable to all samples within the flight hardware, including flight controls and UV-irradiated samples. Spacecraft radiation detection instrumentation (unpublished data) indicated that approximately half of the total Apollo 16 dose originated from the protons and electrons of the trapped radiation (Van Allen) belts during the ascent of the spacecraft through the belts at the time of trans-lunar injection. This dose was received over a relatively short period (30 minutes) at dose rates ranging to a maximum of 2.71 rad hr^{-1}.

4. Conclusions

Solar UV reaching biological test systems in the Apollo 16 microbial response experiment were successfully monitored by the two methods described, allowing for accurate biological dose-response evaluations to be performed. Also high-energy multicharged particle tracks and the integrated ionizing radiation dose were monitored for the flight period. The flight hardware housing the biological test systems was found to have been more shielded than other dosimeter-containing components on the same space flight.

References

[1] G. R. Taylor, Ann. Rev. Microbiol. **28**, 121 (1974).
[2] G. R. Taylor et al., Apollo 16 Preliminary Science Report, NASA Special Publication SP-315, p. 27-11 (1972).
[3] G. R. Taylor et al., BioScience **24**, 505 (1974).
[4] Proc. Symp. on Microbial Response to Space Environment, NASA Tech. Publ. TM X-58103 (1973).
[5] M. Wrighton and S. Witz, Mol. Photochem. **3**, 387 (1972).
[6] E. V. Benton and R. P. Henke, Univ. of San Francisco Contr. Rep. USF-TR-24 (1972).
[7] R. E. Benson and L. S. Pinsky, Apollo 16 Preliminary Science Report, NASA Special Publication SP-315, 27-17 (1972).
[8] H. Bücker, Life Sciences and Space Research XII, 43 (1974).
[9] J. R. Cameron, N. Suntharilingam and G. N. Kenney, Thermoluminescent Dosimetry, Univ. of Wisconsin Press (1968).

* 1 rad = 10^{-2} joule kg^{-1}.

Life Sciences and Space Research XIII — Akademie-Verlag, Berlin 1975

EFFECTS OF SOLAR ULTRAVIOLET RADIATIONS ON *Bacillus subtilis* SPORES AND T_7 BACTERIOPHAGE

J. Spizizen[a], J. E. Isherwood[a] and G. R. Taylor[b]

[a] Scripps Clinic and Research Foundation, La Jolla, Calif., USA
[b] NASA-Johnson Space Center, Houston, Texas, USA

Spores of *Bacillus subtilis* HA 101 and the DNA polymerase I-defective mutant HA 101 (59) F were exposed to selected wavelengths of solar ultraviolet light and space vacuum during the return of Apollo 16. In addition, coliphage T_7 suspensions were exposed to solar ultraviolet radiation as part of the Microbial Response to Space Environment Experiment. Optical filters were employed to provide different energy levels at wavelengths 254 nm and 280 nm. Dose-response curves for lethal and mutagenic effects were compared with ground-based data. A close parallel was observed between the results of solar radiation and ground tests with spores of the two strains. However, significantly greater inactivation of T_7 bacteriophage was observed after exposure to solar ultraviolet radiation.

1. Introduction

The influence of space flight factors on survival and mutation frequencies of a variety of microbial forms has been the subject of numerous investigations [1]. Not only do microbes serve as excellent models of living cells, but an understanding of their survival and potential genetic alterations could be of significance for future space explorations. As a part of the Microbial Response to Space Environment Experiment, flown on the Apollo 16 mission, spores of *Bacillus subtilis* and T_7 bacteriophage were exposed to filtered components of solar ultraviolet light. Spores of two strains of *B. subtilis*, one repair-defective and hence radiation-sensitive, were employed and lethal and mutagenic effects of graded doses of filtered solar ultraviolet determined.

2. Materials and Methods

The design of the complete experiment and a description of the hardware is given in [2]. The strains of *Bacillus subtilis* employed in these studies were HA101 and the DNA polymerase I-deficient strain HA101 (59) F [3]. These strains are characterized by their requirements for L-histidine, L-leucine and L-methionine. Strain (59) F is repair-defective and sensitive to ultraviolet radiation and methyl methane sulfonate. The strains produce characteristic rough colonies on sporulation agar (e.g. AK agar, Baltimore Biological). Mutants altered in colonial morphology including sporulation-defectives are readily detected. The requirement for three amino acids aids in the elimination of contaminants. Spores were prepared on sporulation agar following 48 hours of growth, and purified by lysozyme treatment, Renografin gradient centrifugation and extensive washing in water. Storage in water for one week or longer at 5 °C was required for disaggregation of purified spores.

In ground-based experiments, 1.8 ml aliquots of spore suspensions, not exceeding 10^8 spores per ml were irradiated in small petri dishes. Spore monolayers were also prepared on 0.45 μm Millipore membrane filters (25 mm diameter) by filtration of 10^8 spores, followed by drying at room temperature. For flight studies, aqueous suspensions of 5×10^7 spores per ml were employed in type A cuvettes. Monolayers on Millipore filters were used in types B (unvented) and C (vented) cuvettes. Spores were recovered with the aid of Tween 80 and sonication. T_7 bacteriophage was prepared by lysis of *Escherichia coli* B and purified by centrifugation after refrigeration with polyethylene glycol 20000, followed by cesium chloride density gradient centrifugation. Concentrated stocks were diluted to 1×10^8 plaque-forming units per ml for ground and flight experiments.

Quartz bandpass interference filters with peak wavelengths of 254 and 280 nanometers were employed in ground-based and flight studies. Quartz neutral density filters were employed to titrate the radiant energy. In ground studies a quartz mercury-vapor lamp was used as source of ultraviolet radiation.

3. Ground Based Studies

3.1. Detection of Mutation Frequencies

Previous studies in this laboratory have shown that treatment of spores of *B. subtilis* with the mutagen ethyl methane sulfonate induced a high frequency of asporogenous and colonial morphology mutations [4]. This is due to the fact

Table 1

UV Induced Mutation in Spores of *B. subtilis* 101

Survivors (viable, N/N_0) = 0.03

Mutation to	Mutations/10^5 Survivors	
	Control	4×10^2 ergs mm^{-2} 254 nm
His^+	0.01	8.5
Leu^+	0.005	4.8
$Azide^r$	0.02	9.6
MT^r	0.02	8.5
Morph	50	1.1×10^4
Spo^-	30	5.2×10^3

His^+ = L-histidine prototrophy
Leu^+ = L-leucine prototrophy
$Azide^r$ = azide resistance (100 μg ml)
MT^r = 5-methyl tryptophan resistance (50 μg ml^{-1})
Morph = colonial morphology mutants
Spo- = sporulation-defective mutants

that numerous genes are involved. We have found that highly reproducible data for spontaneous and mutagen-induced mutation frequencies can be obtained using spores under defined conditions. In Table 1 data are presented on the mutant frequencies induced by 254 nm ultraviolet radiation of spores of *B. subtilis* 101. The spontaneous and induced mutation frequencies for the markers *his* (L-histidine), *leu* (L-leucine), *azide* (resistance) and *MT* (5-methyl tryptophan resistance) are similar. By contrast the mutation frequencies for the *morph* (colonial morphology) and *spo-* (sporulation-defective) markers are several orders of magnitude higher. Nevertheless, the induced increases in mutation frequencies are essentially similar.

3.2. Lethal and Mutagenic Effects of UV on Spores

The facility with which the *morph* markers can be detected has been exploited in these studies. Thus, in Fig. 1, are presented data for dose-survival of spores of *B. subtilis* 101 (*w*) and a repair defective strain (*pol-*) irradiated with 254 nm

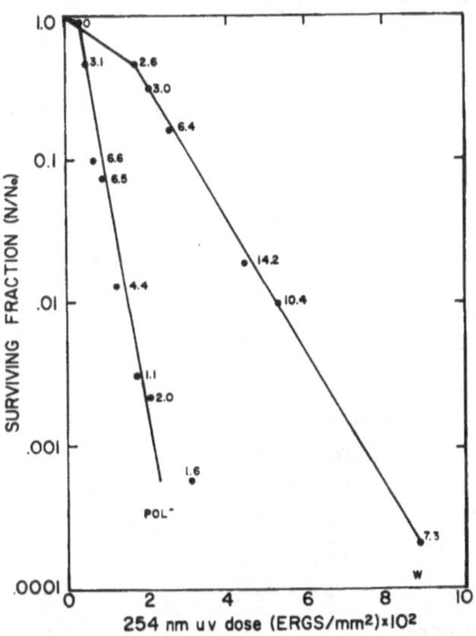

Fig. 1. *B. subtilis* 101 (*w*) and 59F (*pol-*) irradiated with 254 nm UV: ground studies (numbers adjacent to points are frequencies of *morph* mutants, as percentage of survivors).

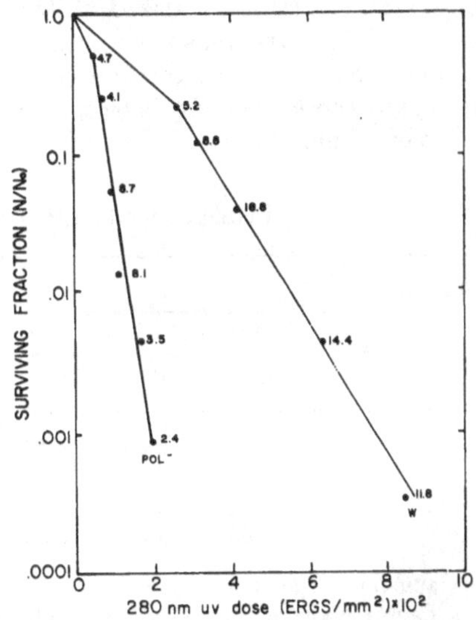

Fig. 2. *B. subtilis*, 101 (*w*) and 59F (*pol-*) irradiated with 280 nm: ground studies (numbers adjacent to points are frequencies of *morph* mutants, as percentage of survivors).

ultraviolet light. The numbers adjacent to the points on the curves represent the mutation frequencies (per cent of survivors) of the *morph* phenotypes. Differences between these strains are evident. The *pol-* strain appears to be less mutable at higher doses of ultraviolet energy, and appears to go through a maximum at doses giving 0.1% survival. By contrast the *w* strain has an increased mutation frequency at higher doses of 254 nm. Similar data have been obtained with 280 nm radiation (Fig. 2), but the *morph* mutation frequencies are significantly higher than those obtained with 254 nm irradiation. With higher wavelengths of ultraviolet light, increased mutation frequencies were observed (unpublished observations in this laboratory).

4. Flight Studies

4.1. Spores and T_7 Coliphage unexposed to Solar Radiation

Studies were made of spores and T_7 phage to determine stabilities, recoveries, effects of vibration and other factors encountered in space flight. Accordingly type A, B and C cuvettes containing spores were held on the ground ("ground" controls), in flight cuvettes, and accompanied the flight hardware at all times during flight. "Vial" controls were held in the authors' laboratory at temperatures similar to the flight hardware. "Vibration" controls were subjected to simulated launch vibrations in flight cuvettes at the NASA Lyndon B. Johnson Space Center at the same time as the lift-off of Apollo 16. "Flight" controls were placed in the flight experiment package but were not exposed to light. The results of these studies are summarized in Table 2. The data are well within experimental error and indicate reasonable recoveries and no effect of flight conditions on viability. Estimates of the frequency of *morph* mutations are given in Table 3 and again there appears to be no apparent effect of flight conditions on the frequencies of mutations.

Table 2

Viable Count Recovered ($\times 10^6$): Unexposed Controls

Location	*B. subtilis*						T_7 Coliphage
	HA 101, spores			HA 101, *pol-*, spores			
	A	B	C	A	B	C	A
Vial	4.30	5.46	6.25	3.77	4.66	4.61	—
Ground	3.78	5.52	5.43	3.17	5.02	4.72	5.61
Vibration	4.12	6.10	5.34	2.95	4.92	4.86	5.45
Flight	3.69	6.18	6.17	3.30	5.14	4.71	4.45

A = aqueous suspensions
B = spore monolayer on filter, unvented, 1 atmosphere
C = spore monolayer on filter, vented

Table 3

Mutant Frequencies: Unexposed Controls

Location	Per cent, *Morph* Mutants					
	HA 101, spores			HA 101, *pol-*, spores		
	A	B	C	A	B	C
Vial	0.03	0.05	0.06	0.08	0.06	0.04
Ground	0.02	0.07	0.04	0.17	0.15	0.14
Vibration	0.02	0.05	0.02	0.09	0.03	0.07
Flight	0.03	0.07	0.06	0.08	0.01	0.19

4.2. Lethal and Mutagenic Effects of Solar Ultraviolet Radiations

More extensive data in flight experiments were obtained with B cuvettes (dry, unvented) due to more reproducible results with such systems. Fig. 3 presents the results obtained with graded doses of 254 nm radiation following a ten-minute exposure to filtered solar radiation. A comparison with ground-based data is also presented in this figure. The data indicate slightly greater resistance to the lethal effects of 254 nm in flight in the two strains. In addition, lower frequencies of *morph* mutations were observed in the flight experiment compared

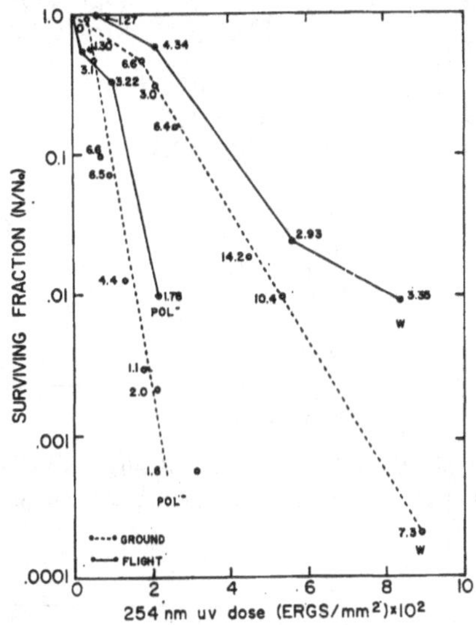

Fig. 3. 254 nm irradiation of *B. subtilis* 101 (*w*) and 59F (*pol-*): ground and flight studies (numbers adjacent to points are frequencies of *morph* mutants, as percentage of survivors).

with ground data. This may not be of significance since cloning of the flight experiments was limited relative to ground studies. Less correspondence of the lethal effects of 280 nm was observed in the flight study compared with ground data (Fig. 4). Spores of both strains appeared to be less susceptible at all energy levels in the flight study compared with ground irradiation. Mutation frequencies in the flight tests appear to be reasonable for this range of radiation energies.

The data for type C cuvettes (dry, vented to solar vacuum) were limited, and only a few energy levels investigated. It appeared that the 101 (*w*) strain was more susceptible than under conditions of 1 atmosphere, both with 254 nm and 280 nm radiations. However, the 59F (*pol-*) strain appeared to be as sensitive as in the B cuvette samples. Mutation frequencies were similar in the samples from both types of cuvettes.

The inactivation of T_7 bacteriophage by graded doses of 254 nm radiation in ground and flight experiments, the latter in type A cuvettes (aqueous suspension), is shown in Fig. 5. The flight and ground data show comparable biphasic curves. However, it appears that T_7 was more susceptible to inactivation in flight, compared with ground studies. It should be recalled that unexposed controls were identical with ground controls.

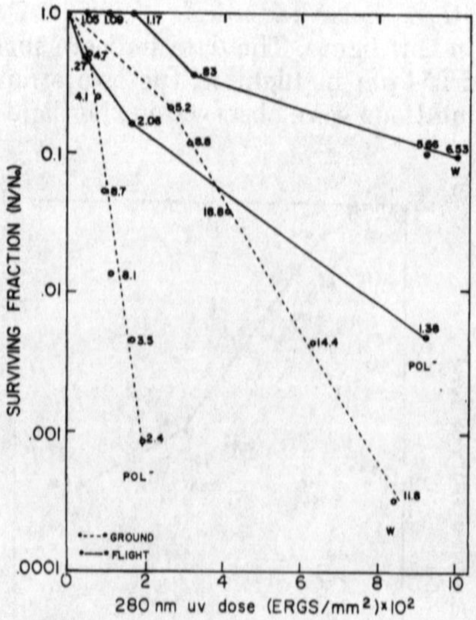

Fig. 4. 280 nm irradiation of *B. subtilis* 101 (*w*) and 59F (*pol-*): ground and flight studies (numbers adjacent to points are frequencies of *morph* mutants, as percentage of survivors).

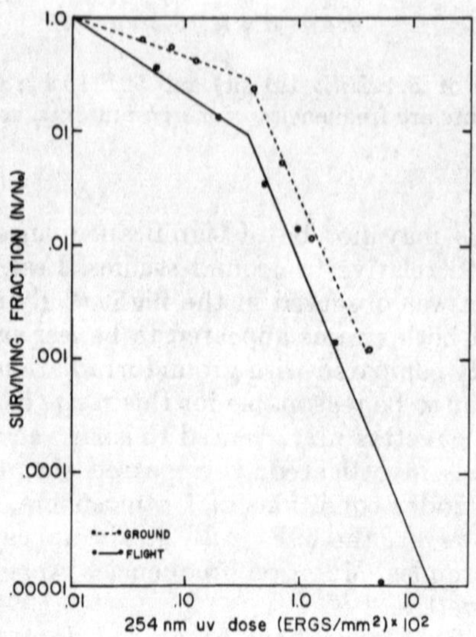

Fig. 5. T_7 bacteriophage inactivation by 254 nm ultraviolet radiation.

5. Conclusions

1. It is evident that spores of *Bacillus subtilis* can be satisfactorily used as monitors of lethal and mutagenic effects of space environments.

2. Under conditions of the present study in the absence of solar irradiation, no significant effect of flight conditions was observed on survival and mutation frequency of the strains used.

3. Filtered solar radiation at 254 nm and 280 nm appeared to be less lethal to spores at the doses employed and induced lower mutation frequencies than comparable ground studies.

4. By contrast, T_7 bacteriophage was inactivated to a greater degree by 254 nm solar ultraviolet than in ground studies, although the shapes of the dose-response curves were comparable.

5. Space vacuum appeared to enhance the lethal effect of solar ultraviolet radiation both at 254 nm and 280 nm with the wild type strain but not with the repair-defective strain of *B. subtilis*. However, no difference in mutation induction was noted with either strain compared with irradiation in 1 atmosphere.

References

[1] G. R. Taylor, Ann Rev. Microbiol. **28**, 121 (1974).
[2] G. R. Taylor et al., Apollo 16 Preliminary Science Rep., NASA Special Publication SP-315 (1972).
[3] K. G. Gass et al., J. Bacteriol. **108**, 364 (1971).
[4] J. Ito et al., Mutation Res. **13**, 92 (1971).

Life Sciences and Space Research XIII — Akademie-Verlag, Berlin 1975

CALCULATED FREQUENCY OF COSMIC RAY HEAVY NUCLEI ENDERS INSIDE A SPACE SHIP

O. C. ALLKOFER[a] and W. HEINRICH[b]

[a] Institut für Reine und Angewandte Kernphysik, Universität Kiel, Kiel, FRG
[b] Gesamthochschule, Siegen, FRG

As has been described earlier, the interaction parameters of heavy nuclei in air were determined by comparison of calculated and measured fluxes. The frequency of stopping heavy nuclei in absorber depths of $0-40$ g cm^{-2} for the periods of maximum and minimum solar activity were calculated using these parameters. The results are presented for four charge groups ($6 \leqq Z \leqq 9$, $10 \leqq Z \leqq 14$, $15 \leqq Z \leqq 19$, $20 \leqq Z \leqq 28$). The calculated frequencies for solar maximum are compared with measured fluxes inside the Biostack 1, flown on board Apollo 16.

A full paper is published in the *Journal of Spacecraft and Rockets.*

Life Sciences and Space Research XIII — Akademie-Verlag, Berlin 1975

RADIOBIOLOGICAL RESULTS OF THE BIOSTACK EXPERIMENT ON BOARD APOLLO 16 AND 17

E. H. GRAUL[a], W. RÜTHER[a], W. HEINRICH[b], O. C. ALLKOFER[b], R. KAISER[c], R. PFOHL[c], E. SCHOPPER[d], G. HENIG[d], J.-U. SCHOTT[d] and H. BÜCKER[e]

[a] Klinik und Poliklinik für Nuklearmedizin, Universität Marburg (Lahn), FRG
[b] Institut für Reine und Angewandte Kernphysik, Universität Kiel, Kiel, FRG
[c] Centre de Recherches Nucleaires, Strasbourg-Cronenbourg, France
[d] Institut für Kernphysik, Universität Frankfurt, Frankfurt (Main), FRG
[e] Arbeitsgruppe für biophysikalische Weltraumforschung, Universität Frankfurt, Frankfurt (Main), FRG

After penetrating the Biostack capsule, some of the HZE particles hit the biological objects carried: bacterial spores (*Bacillus subtilis*), seeds (*Arabidopsis thaliana* and *Vicia faba*), and shrimp eggs (*Artemia salina*). The different biological objects were affected by heavy ions in widely varying ways. A broad range of radiobiological investigations has been carried out in regard to the objects' response to HZE particles.

The most sensitive biological objects in the Biostack experiments proved to be the shrimp eggs. The development of 500 eggs hit by heavy cosmic ions was investigated. This differed significantly from the flight controls (eggs flown in the Biostack but not hit by heavy ions) and from the ground controls. From this it has been concluded that penetration on the part of a single heavy ion may injure the encysted blastula. This damage was found to influence gastrula formation and even the hatching process of the nauplius. Abnormalities (increased by a factor of 10) in the orthonauplius were observed during the development of the hit eggs; they consisted, for example, of shortened extremities or an abnormal thorax or abdomen.

In addition, eggs of *Tribolium confusum* and *Carausius morosus*, which were included in Biostack 2 (Apollo 17), have been investigated, and the influence of single heavy ions on the development process of these highly organized insects has been studied.

1. Introduction

The Biostack experiment (Principal Investigator H. Bücker) was exposed to cosmic galactic radiation. It is therefore possible to study the combined action of individual heavy nuclei of the cosmic radiation and space flight factors on biological systems in the resting state.

Direct experimental evidence of the effect of individual heavy particles was presented by Chase [1] in the form of greying of hair in the coats of balloon-borne mice, by Brustad [2] on maize embryos, by Yagoda et al. [3] and Haymaker et al. [4] on brain injury in balloon-borne monkeys, and by Eugster [5] on cellular destruction in *Artemia salina* eggs in balloon flights.

2. Materials and Methods

The objects were selected on the basis of the following considerations: (i) they had to endure exposure in contact with the track detector, and embedding in PVA; (ii) they should comprise a variety of biological specimens, such as animal eggs, bacterial spores, and plant seeds, in order to study radiation effects on different levels of biological organization; (iii) they should be radiosensitive and perhaps also show effects of other flight factors.

The first investigations were made on *Artemia salina*. This belongs to the order Branchiopoda, sub-order Euphyllopoda, and lives in salt water [6–8]. Its life

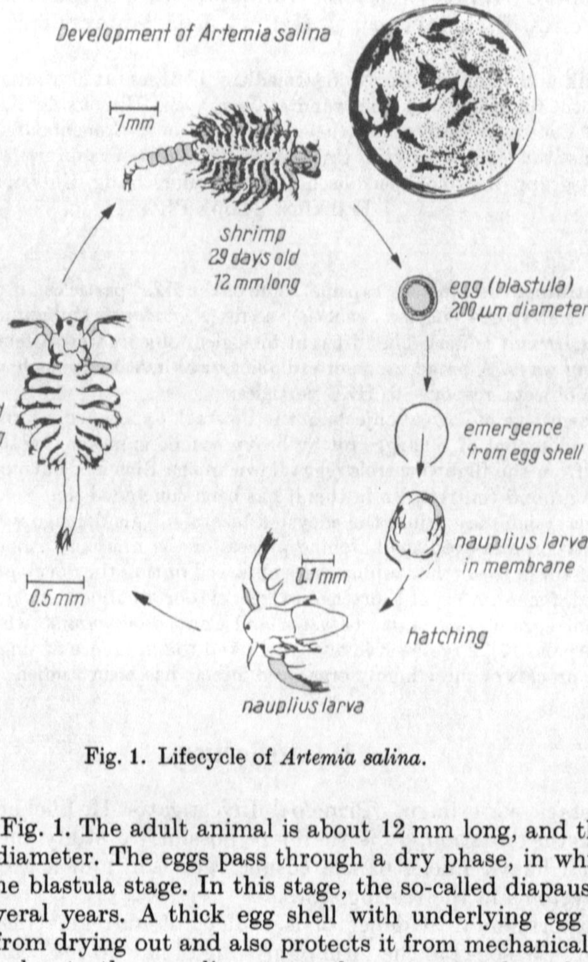

Fig. 1. Lifecycle of *Artemia salina*.

cycle is shown in Fig. 1. The adult animal is about 12 mm long, and the eggs are about 200 µm in diameter. The eggs pass through a dry phase, in which the egg cells develop to the blastula stage. In this stage, the so-called diapause, the eggs are viable for several years. A thick egg shell with underlying egg membrane prevents the egg from drying out and also protects it from mechanical injury.

In order to develop to the nauplius stage, the eggs must be cultivated in a special saline solution. At a temperature of about 20°–25 °C the nauplius develops from the egg in between 20 and 48 hours. Its food requirements can easily be satisfied by the alga *Dunaliella viridis*, which is added to the *Artemia salina* cultures in concentrated form. Thus it is possible to cultivate single shrimps or large numbers of them over their life cycle of about 9 months. Under these conditions the female nauplii grow to a length of about 12 mm, and the male to about 10 mm.

The plates in the experiment consisted of a monolayer of eggs and comprised (i) those flown but not hit by heavy particles (900); (ii) those flown and hit by heavy particles (500); (iii) those not flown and not hit, the ground controls (1400).

The individual eggs hit by heavy primaries were picked up in relation to a coordinate grid (Fig. 2) under a microscope, and cultivated in small glass dishes. Observation began after 12 hours of cultivation. We investigated the development of about 500 eggs hit by heavy cosmic ions. This differed significantly from the flight controls and from the ground controls.

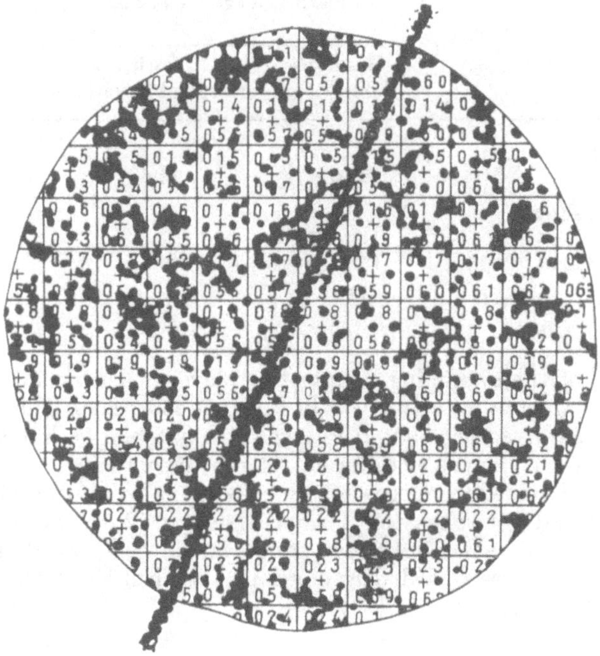

Fig. 2. *Artemia salina* eggs in PVA-foil with underlying coordinate grid showing the track of a heavy primary (photomicrograph).

3. Results

The eggs that were hit (Table 1) have a low hatching rate (10.28%, development up to the swimming nauplius stage) compared with ground-control eggs (89.8%). Likewise, the survival rate of these nauplii was low, since only a few of them grew to adults in about four weeks. But the eggs that were not hit but received background radiation and were subject to the flight environment also showed a reduced rate of hatching (46.2%) compared with the ground controls (89.87%). The left column of Table 1 lists the numbers of eggs that showed no development, which was high for those hit by heavy ions. But a smaller number of eggs that received background radiation also showed no development. The heavy ions damaged the blastula in the diapause stage. Depending on the variable local radiation damage to the individual blastula cells, only a small number of larvae developed into swimming nauplii during subsequent development, as is shown in the centre column of Table 1.

Table 1
Frequency (%) of different Developmental Stages after Irradiation Damage by Heavy Primaries and at Controls

	Eggs not developing	Eggs showing development (egg shell at least cracked)		
		No tearing of egg membrane		Tearing of egg membrane
		Basic shape of larva not formed	Basic shape of larva formed	Swimmung nauplius larva
Eggs hit by heavy primaries	45.0	29.16	15.56	10.28
Eggs not hit by heavy primaries; background irradiated	21.75	15.0	17.0	46.25
Ground controls	9.6	0	0.6	89.8

The vitality of the swimming nauplii that developed from hit eggs was also low. A high death rate was observed in the first four days. The growth of the animals which survived the first few days after hatching is indicated in Fig. 3 by the length of the animals at different times after hatching.

Larvae which hatched from eggs hit by heavy ions were stunted (about 6 mm in length), and did not reach the normal length of 12 mm even 25 days after hatching. The nauplii which developed from eggs flown in the Biostack, but not hit by heavy ions, were retarded in development until the 13th day, but then grew to normal length.

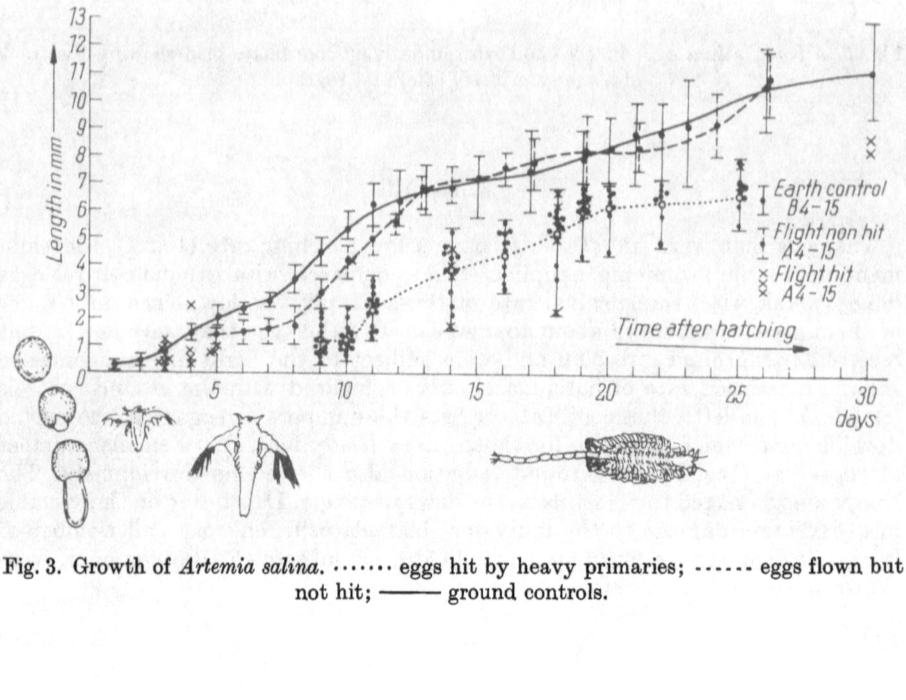

Fig. 3. Growth of *Artemia salina*. ······· eggs hit by heavy primaries; ------ eggs flown but not hit; ——— ground controls.

Table 2
Artemia salina from Apollo 16: Development and Reproduction
Sal. = 3%, T = 25 °C

	Sexual maturity*	Days to egg-production	Number of nauplii**	Number of nauplii (F_1)***
Ground control	8–13	5– 9	(5) 132–*148*–160	505–*590*–645
Flown but not hit	9–14	5–10	(3) 105–*126*–148	225–*251*–287
Flown and hit	12–24	20–24	(0–2) 0– *82*–112	0–*102*–169

Salinity = 3% NaCl. Temperature 25 °C. The values show the ranges; means given in italics.

*Days after hatching to reach sexual maturity (mating).

** Number of nauplii per brood and per female obtained in two months (number of broods in parentheses).

*** Number of nauplii in two months in the F_1 generation per female.

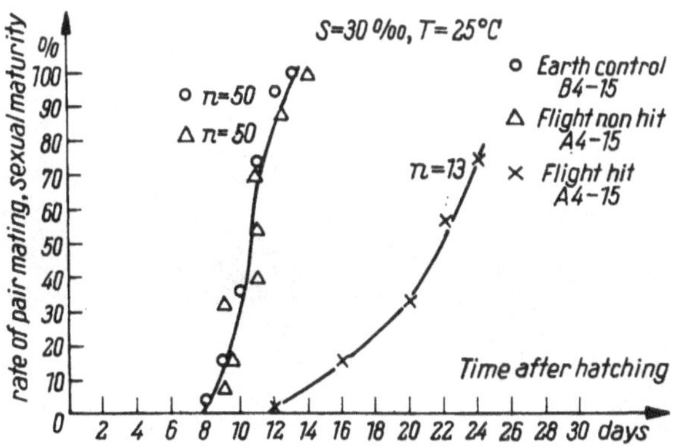

Fig. 4. Sexual maturity of *Artemia salina* (pair matings). ×, eggs hit by heavy primaries; Δ, eggs flown but not hit; ○, ground controls.

Shrimps from eggs that were hit reached maturity relatively late (Table 2 and Fig. 4). The number of descendants in the F_1 generation of individual pairs was examined over a period of 60 days, and, as shown in Table 2, the number of nauplii from pairs which had hatched from flown eggs was reduced, especially if the eggs had been hit by heavy primaries.

Anomalies of the body and extremities (Figs. 5, 6, 7) were observed more frequently in the nauplii which had developed from eggs hit by heavy primaries; they occurred in 6% of the nauplii. In the other groups such abnormalities were relatively rare (0.5%).

4. Conclusion

In summary, there was a significant difference in the development of nauplii from *Artemia salina* eggs hit by heavy ions, eggs which received background irradiation in flight, and ground control eggs. A single heavy ion may injure the

Fig. 5. A normal nauplius (left) compared with an abnormality of development of the abdominal segment (right).

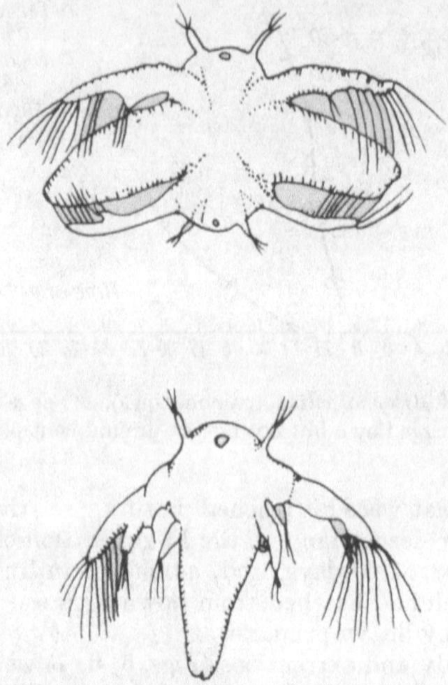

Fig. 6. Abnormal nauplius (with two abdomina and two heads) of *Artemia salina* hatched from an egg that had been hit by an HZE particle compared with a normal nauplius.

encysted blastula. This damage was found to influence gastrula formation, and even the hatching process, thus suggesting that the target involved in inhibition of hatching must be very small, and may require an enormous but highly localized dose of radiation.

In addition, eggs of *Tribolium confusum* and of *Carausius morosus*, which were

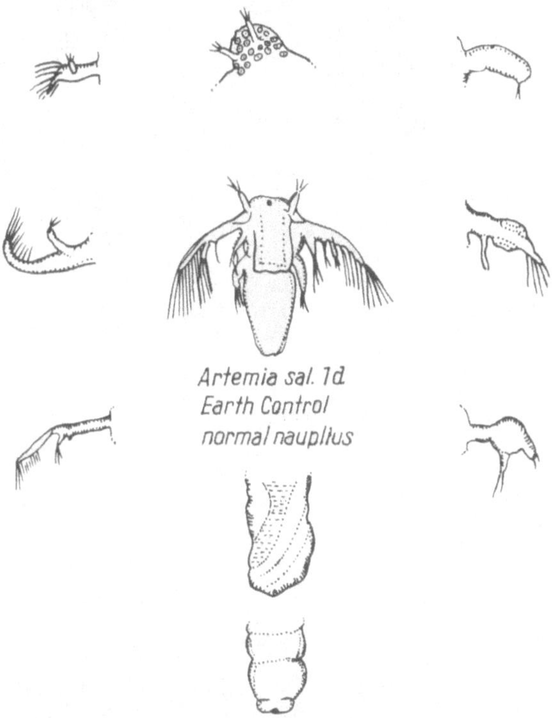

Fig. 7. Body abnormalities of *Artemia salina* nauplii hatched from hit eggs compared with a normal nauplius.

included in Biostack 2 (Apollo 17) were investigated. Here, the influence of a single heavy ion also damages the development process and leads to similar results: a low hatching rate, body abnormalities, low vitality, and a short life-span. These organisms can tolerate a certain amount of injuries induced by low LET radiation without any significant alteration in their functional state, but cannot tolerate high LET heavy-ion injury. Here, a single hit can cause irreversible damage to the organism.

References

[1] H. B. Chase, J. Aviat. Med. **25**, 388 (1954).
[3] T. Brustad, Rad. Res. **14**, 139 (1961).
[4] H. Yagoda et al., Military Med. **128**, 655 (1953).
[5] W. Haymaker et al., Aerospace Med. **41**, 989 (1970).
[6] J. Eugster, Weltraumstrahlung Medizin, Verlag Hans Huber, Bern, Switzerland 1955.
[7] D. M. Whitaker, J. Exp. Zool. **83**, 391 (1940).
[8] T. Iwasaki, J. Radiat. Res. **5**, 69 (1964).
[9] Y. H. Nakanishi, I. Iwasaki, T. Okigaki and H. Kato, Proc. Jap. Acad. **39**, 306 (1963).

Life Sciences and Space Research XIII — Akademie-Verlag, Berlin 1975

RESULTS OF THE *Bacillus subtilis* UNIT OF THE BIOSTACK II EXPERIMENT: PHYSICAL CHARACTERISTICS AND BIOLOGICAL EFFECTS OF INDIVIDUAL COSMIC HZE PARTICLES

H. Bücker, R. Facius, D. Hildebrand and G. Horneck

Arbeitsgruppe für biophysikalische Weltraumforschung, Universität Frankfurt, Frankfurt (Main), FRG

The effectiveness of cosmic HZE-particles on unicellular procaryotic organisms was studied on *Bacillus subtilis* spores, which were accommodated in the Biostack I and II experiments on board Apollo 16 and 17. Identification of the spores that were hit was achieved by using the Biostack sandwich construction and by precise microscopical measurements of tracks of particles. Germination, outgrowth and the rate of cellular elongation were investigated. A method was developed to determine the charge of each individual HZE particle that penetrated a spore and its energy loss in the region of hit. An attempt was made to establish a connection between these physical characteristics and the biological effects produced.

1. Introduction

Studies on the biological efficiency of heavy primaries of cosmic radiation, or HZE particles [1], require precise determination of both the biological region which has been penetrated by an HZE particle and the physical characteristics of the particle. This has been achieved in the Biostack experiments flown on board the Apollo 16 and 17 spacecraft. The experiment has a sandwich construction, with monolayers of different dormant biological systems (spores, seeds and eggs) between specially selected physical track detectors and this was repeated in all biological sub-units of the stack [2]. Different methods were used to localize precisely the point of interaction of an HZE particle in a biological layer. The accuracy in locating this point attained ± 1 µm. For animal eggs or plant seeds, which all exceeded 50 µm in diameter, this accuracy was sufficiently high to detect even the region that was hit inside the biological specimen. In the case of bacterial spores, however, which were about 1.5 µm in length, this current limit of spatial resolution meant that usually several spores hat to be considered as possibly being hit by the cosmic particle. Accordingly all spores within the "target area", a region of about 12 µm^2 around the expected point of penetration, had to be taken into consideration in the biological investigations. The spores outside this target area were treated as flight controls and considered to have a probability of less than 0.3% of comprising a spore that had been hit by the cosmic particle.

2. Results and Discussion

2.1. Development of the Spores inside the Target Area

The methods of post-flight evaluation of the *Bacillus subtilis* unit of Biostack II were similar to those described for the Biostack I analysis [3]. A more radiosensitive polymerase negative mutant (pol^-) of *Bacillus subtilis* was used in addition to *Bacillus subtilis* 168. The biological processes under investigation were germination and outgrowth of the spores, and the elongation of the first vegetative cell. In these first developmental steps, no difference was detected between the

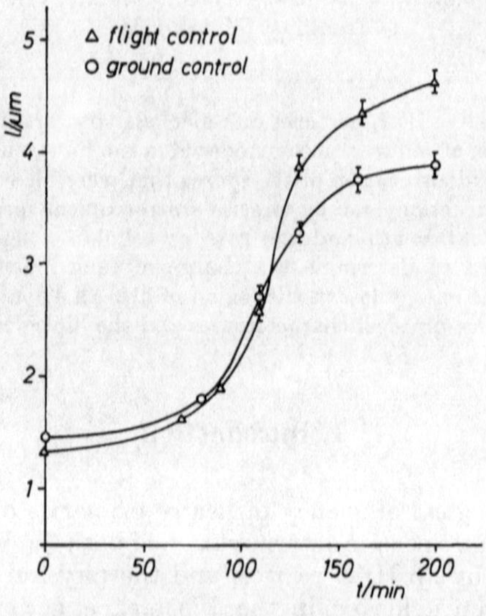

Fig. 1. Elongation of the first vegetative cell, outgrown from *Bacillus subtilis* 168 ground control and flight control spores.

ground controls and the flight controls (Fig. 1). This indicates the resistance of spore development to the additional inevitable factors involved in space flight, as observed earlier in the Biostack I studies [3]. Thus, in the following studies on the effects of HZE particles, the influence of the other environmental factors of space flight could be neglected.

The development of the spores lying inside the target area of a cosmic particle also confirmed the results of Biostack I [3]. All of the spores germinated (Fig. 2). However, the fraction that gave outgrowth was significantly reduced compared with the flight controls (Fig. 2). The development was observed over a three-hour incubation period. Sometimes, only swelling of the spores was found (Fig. 3), which is not caused by active growth and metabolism. These results indicate that HZE particles have a damaging effect on the outgrowth of bacterial spores. A more detailed multivariate analysis is in progress, but preliminary results seem to confirm the simple comparison of mean values.

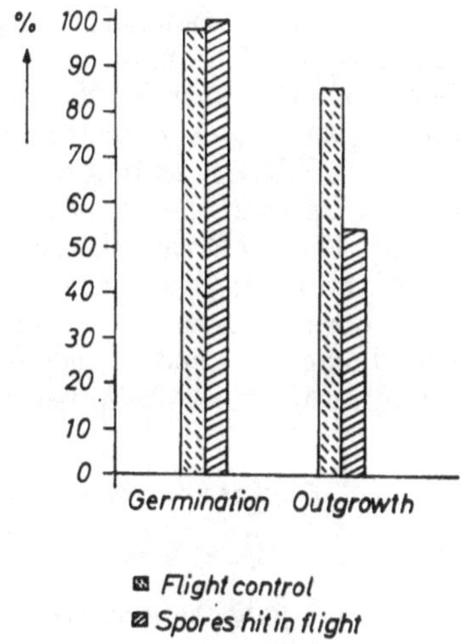

Fig. 2. Germination and outgrowth of *Bacillus subtilis* pol⁻ spores lying inside the target area compared with the flight controls.

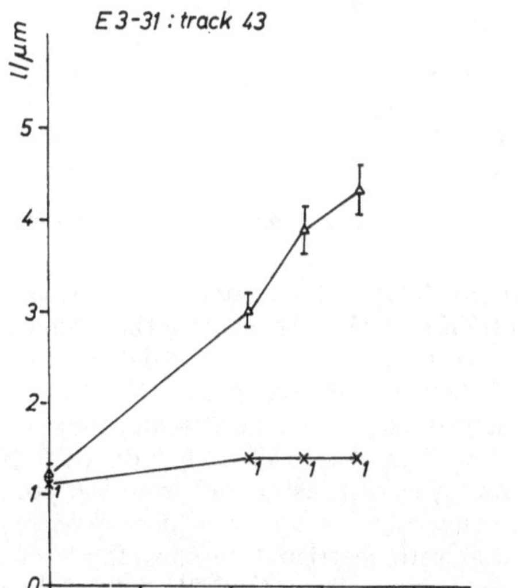

Fig. 3. Growth inhibition of a *Bacillus subtilis* pol⁻ spore (x_1) lying in the target area of a particle. Cellular elongation of the first vegetative cells outgrown from the adjacent flight control spores (△).

2.2. Physical Data on Tracks producing a Hit in the Spore Layer

The sandwich construction of Biostack made it possible to determine the physical characteristics of those HZE particles that caused a biological effect. The physical quantities of main interest are the charge, the energy and the energy loss of the particles. The correlation of these parameters with the biological effects will contribute important information on the microdosimetric approach to radiation hazards of HZE particles. The knowledge of the energy loss in the hit region alone is not sufficient for this purpose. This may be illustrated by the different response of different visual track detectors to the penetration of an HZE particle. Furthermore the total dose is an improper measure of the biological effectiveness of HZE particles. The total dose was approximately 700 mrad for the 12-day mission of Apollo 17. The HZE particle dose, calculated from nuclear emulsion data for particles with an energy loss exceeding 80 keV μm^{-1}, was only 3.5 mrad, i.e. 0.5% of the total dose of cosmic radiation. However, its deleterious influence on the outgrowth of bacterial spores has been detected and many significant effects on other biological objects have been proved by means of the

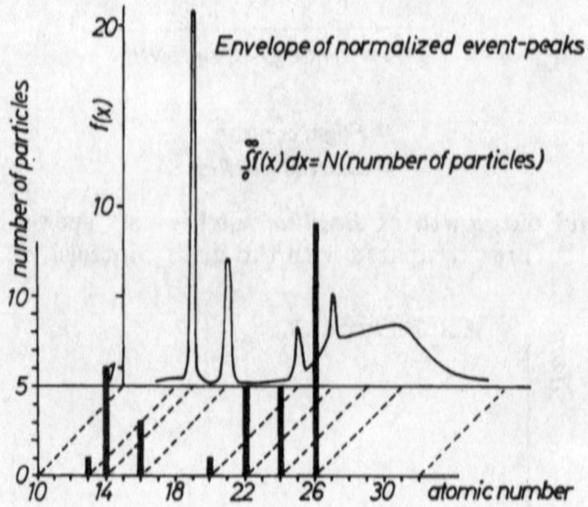

Fig. 4. Frequency of atomic numbers of hitting particles in the *Bacillus subtilis* unit.

Biostack method. This invalidates the common dose concept as a measure of the biological efficiency of HZE particles. An effect of the cosmic background radiation on the *Bacillus subtilis* spores, however, was not detected.

The physical data of those tracks that produced a hit in the spore layer were determined by the analysis of the plastic detector sheets, which were stacked between the spore layers. The charge of each individual HZE particle and its energy loss in the spore layer were estimated from the profiles of the etch cone lengths in the different sheets.

The biological studies were restricted to the effects of HZE particles with $Z > 12$ and a restricted energy loss [4] REL $\geqq$ 1.2 GeV cm^2 g^{-1}. Up to now, only a little information exists on their biological efficiency, because the availability of particles of very high charge and a very high energy loss over an extended track length by experimental equipments on earth is limited. The particles which produced a hit in the spore layer belong to an interval from silicon up to the iron group (Fig. 4). Their REL reached values up to 5.0 GeV cm^2 g^{-1} (Fig. 5). It is not yet possible to correlate the measured biological effects on the spores that were hit with the physical parameters of the penetrating HZE particles. Only slight evidence is available in our data for the biological effects being a function of the REL.

The two-fold angular distribution of all penetrating particles has also been determined (Fig. 6). Its deviation from the distribution theoretically expected indicates a marked effect of the absorption of the cosmic rays by the material surrounding the Biostack. A comparison with the corresponding results of Biostack I [3] also confirms this. In Apollo 16 the *Bacillus subtilis* unit was placed pointing to the interior of the command module. This resulted in two

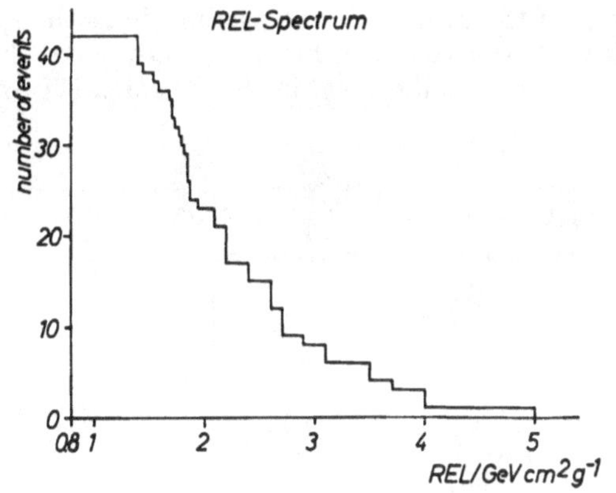

Fig. 5. Integral REL-distribution of hit events in the *Bacillus subtilis* unit.

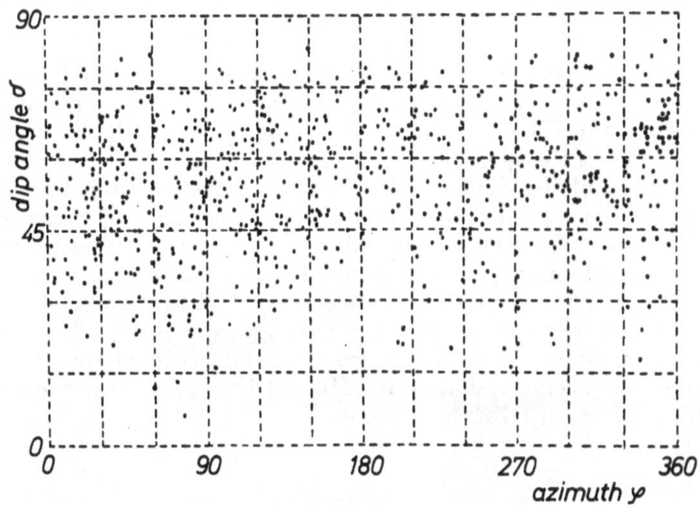

Fig. 6. Angular distribution of HZE particles in the *Bacillus subtilis* unit.

distinct, and nearly opposite, areas of reduced fluence. In Apollo 17, however, the *Bacillus subtilis* unit was fixed next to the wall of the spaceship. As a consequence the two regions merged, giving now one broad region of absorption at low dip angles. In future experiments the azimuthal orientation of the Biostack will be recorded. Together with the known mass distribution these experimental data may be useful for estimating the influence of matter on the fluence and spectrum of the cosmic radiation.

3. Concluding Remarks

The studies on the effects of HZE particles on *Bacillus subtilis* spores were limited by the small size of these unicellular organisms. Because the hit area of an HZE particle was of the same order of magnitude as the spore length, several spores had a significant probability of being the individual that was hit for each track. Therefore, in further studies, e.g. in the Biostack III experiment, which is

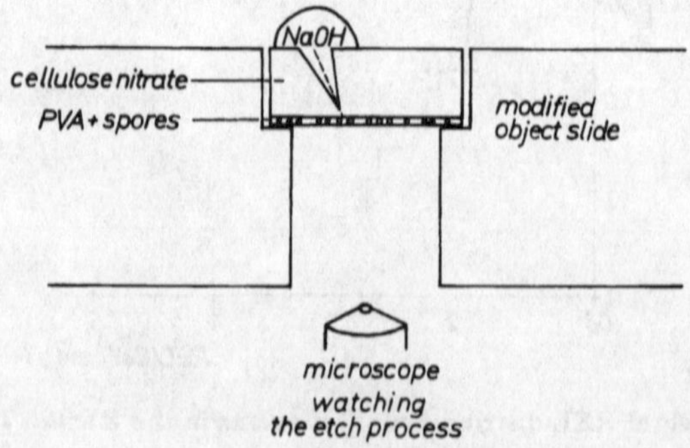

Fig. 7. Microetching, a precise method to determine the biological object hit (to be developed for Biostack III).

scheduled for the Apollo-Soyuz mission, it is planned to improve the method of determination of the spore hit. This may be achieved by individual microetching of each track in the CN sheet which carries the spores, until the tip of the etch cone nearly touches the spore layer (Fig. 7).

The sensitivity of the spores to HZE particles is shown by the reduced outgrowth of the spores. This illustrates one manifestation of damage by HZE particles. However, mutation or other damage, that only becomes manifest at a later step of development, might be possible. To study such phenomena it is necessary to isolate the spore that has been hit, which is a planned procedure for the Biostack III experiment system. Such isolated spores will allow investigation of the development up to the macrocolony, of metabolic alterations and of the production of mutations, and Biostack III will thus give further information on the biological effects of HZE particles.

References

[1] D. Grahn, Ed., Report of the National Academy of Sciences, Washington, D. C. (1973).
[2] H. Bücker et al., Life Sciences and Space Research XI, 295 (1973).
[3] G. Horneck, W. Enge, R. Beaujean and K. P. Bartholomä, Life Sciences and Space Research XII, 75 (1974).
[4] E. V. Benton, Report USNRDL-TR-68-14 (1968).

Life Sciences and Space Research XIII — Akademie-Verlag, Berlin 1975

MODIFYING EFFECT OF DYNAMIC SPACE FLIGHT FACTORS ON RADIATION DAMAGE OF AIR-DRY SEEDS OF *Crepis capillaris* (L) WALLR

E. N. VAULINA and L. N. KOSTINA

Academy of Sciences of the USSR, Moscow, USSR

The influence of dynamic factors (vibration and linear acceleration) on the rate of chromosome aberrations in *Crepis capillaris* was studied. The vibrational process simulated was similar in its characteristics to that occurring at the launch of spaceships. In combination with linear acceleration it caused a statistically significant increase in the rate of chromosome aberrations compared with the control ($R = 7.70$).

The dynamic factors modified the effect of radiation damage induced by acute γ-irradiation (3 krad). Pre-radiation treatment with vibration and acceleration on the seeds caused a significant decrease ($R = 10.23$) of the effect of radiation damage, from 15.57% to 9.74%.

The post-radiation treatment of *C. capillaris* seeds with the dynamic factors did not change the rate of chromosome aberrations significantly (from 15.57% to 15.90%).

The earlier experiments carried out on the automatic station Zond 8, spaceship Soyuz 9 and orbital station Salyut have shown that space flight factors increase the rate of chromosome aberrations in non-irradiated seeds. A prolonged action of space flight factors increases the effect of the pre-flight irradiation of seeds and does not affect the result of their post-flight irradiation (Tables 1, 2) [1].

Experiments on *Tradescantia* microspores [2, 3] and mouse marrow cells [4—8] have shown that dynamic space flight factors can increase the rate of chromosome aberrations and modify the radiation damage. In order to determine the role of dynamic space flight factors and their effect on the hereditary structure of plant cells the effect of vibration and linear acceleration in combination with γ-irradiation on air-dry seeds *Crepis capillaris* was studied. In order to reproduce the vibrational conditions during space flights as correctly as possible a wideband random process was constructed and recorded with the maximum power-spectrum density (up to $0.1\,g^2\,\mathrm{Hz}^{-1}$) of vibration in the frequency range of 300—800 Hz and total average quadratic level of about $7-10\,g$. The character of vibrational changes over time was similar to that at the launch of a spaceship. The recorded vibration was reproduced with the help of a tape recorder connected to the table of an electrodynamic vibrator VEDS-10-A. The seeds were also exposed to linear acceleration. This was done on a special centrifuge installed on the vibrator [9]. The vectors of vibrational and linear accelerations were mutually perpendicular. Linear acceleration amounted to $6-8\,g$. *Crepis capillaris* seeds were put into test-tubes in a thin layer and pressed, and after that the test-tubes were fixed on the vibrational centrifuge. The seeds were exposed to the dynamic space flight factors for 9 minutes.

Table 1

Effects of Space Flight Factors on Air-dry Seeds of *Crepis capillaris* (L) Wallr

Variants	Zond 8				Soyuz 9				Salyut			
	No. studied*	Cells with aberrations		R	No. studied*	Cells with aberrations		R	No. studied*	Cells with aberrations		R
		abs. no.	$\% \pm m$			abs. no.	$\% \pm m$			abs. no.	$\% \pm m$	
Control	2283	—	—		22937	35	0.15 ± 0.03	5.37	2819	22	0.78 ± 0.16	1.17
Flight	2398	4	0.17 ± 0.08		8742	54	0.62 ± 0.08		4233	45	1.06 ± 0.16	
γ_1	1274	50	3.92 ± 0.54	0.62	4785	420	8.78 ± 0.41	0.90	1599	687	42.96 ± 1.24	7.40
γ_1 + flight	4071	144	3.54 ± 0.29		5925	550	9.28 ± 0.38		1017	586	57.62 ± 1.55	
γ_2	3954	173	4.38 ± 0.33	0.42	3822	77	2.01 ± 0.23	2.41	2165	70	3.23 ± 0.38	0.02
Flight + γ_2	2322	107	4.61 ± 0.43		5207	70	1.34 ± 0.16		2918	94	3.22 ± 0.33	

* No. studied = no. of metaphases studied.

m is a standard variation given by $m = \left(\frac{\%(100-\%)}{n}\right)$ where

$\%$ = percentage of cells with chromosome rearrangements and
n = number of analysed cells

R is the significance of the difference between experimental and control variants estimated by:

$$R = \frac{\overline{X}_1 - \overline{X}_2}{(m_1^2 + m_2^2)^{1/2}}$$ where

$\overline{X}$ = arithmetic mean, and m is defined above.

Student's criterion is used for $R \geq 3$ and the level of significance $p > 0.01$.

Table 2

Cytological Analysis of *Crepis capillaris* Seeds

Variants	Zond 8 Aberrations						Soyuz 9 Aberrations						Salyut Aberrations					
	(1)		(2)		(3)		(1)		(2)		(3)		(1)		(2)		(3)	
	abs. no.	$\% \pm m$	abs. no.	$\% \pm m$	abs. no.	$\% \pm m$	abs. no.	$\% \pm m$	abs. no.	$\% \pm m$	abs. no.	$\% \pm m$	abs. no.	$\% \pm m$	abs. no.	$\% \pm m$	abs. no.	$\% \pm m$
Control	—	—	—	—	—	—	11	0.05 ±0.04	10	0,04 ±0.04	14	0.06 ±0.05	8	0.28 ±0.108	3	0.11 ±0.23	11	0.39 ±0.129
Flight	2	0.08 ±0.054	1	0.042 ±0.047	1	0.042 ±0.047	22	0.25 ±0.10	14	0.16 ±0.085	18	0.21 ±0.095	5	0.12 ±0.23	10	0.24 ±0.10	28	0.66 ±0.166
γ_1	13	1.02 ±0.281	4	0.31 ±0.154	33	2.59 ±0.44	94	0.20 ±0.125	15	0.31 ±0.154	311	6.50 ±0.69	239	14.95 ±0.99	9	0.56 ±0.209	439	27.45 ±1.24
γ_1 + flight	32	0.79 ±1.37	6	0.15 ±0.065	106	2.60 ±0.248	154	2.60 ±0.248	15	0.25 ±0.08	381	6.34 ±0.384	260	25.57 ±0.682	5	0.49 ±0.10	321	31.56 ±0.733
γ_2	66	1.67 ±0.204	15	0.38 ±0.10	92	2.33 ±0.24	21	0.55 ±0.12	10	0.26 ±0.085	46	1.20 ±0.058	22	1.02 ±0.161	1	0.05 ±0.031	47	2.17 ±0.23
Flight + γ_2	29	1.25 ±0.23	6	0.26 ±0.10	72	3.10 ±0.359	26	0.50 ±0.145	10	0.19 ±0.09	34	0.65 ±0.166	26	0.89 ±0.194	3	0.10 ±0.63	65	2.23 ±0.306

(1) isolocus breaks (NUpd); (2) chromatid; (3) chromosal.

Table 3

Cytological Analysis of *Crepis capillaris* Seeds after the action of Dynamic Factors (vibration + acceleration) and γ-irradiation

Conditions	No. of metaphases	No. of cells with aberrations		R	Aberrations					
					isolocus breaks (NUpd)		chromatid		chromosomal	
		Absolute number	Mean % $\pm m$		Absolute number	Mean % $\pm m$	Absolute number	Mean % $\pm m$	Absolute number	Mean % $\pm m$
Control	4969	15	0.30 ± 0.10		6	0.12 ± 0.04	5	0.10 ± 0.04	3	0.06 ± 0.11
γ	6524	1016	15.57 ± 0.45		200	3.07 ± 0.22	31	0.48 ± 0.10	573	8.78 ± 0.34
v + a	5172	95	1.84 ± 0.17	7.70*	32	0.62 ± 0.10	13	0.25 ± 0.22	62	1.20 ± 0.15
v + a followed by γ	7004	682	9.74 ± 0.36	10.23	172	2.46 ± 0.17	37	0.53 ± 0.10	523	7.47 ± 0.32
γ followed by v + a	7170	1140	15.90 ± 0.44	0.53	276	3.85 ± 0.22	38	0.53 ± 0.10	761	10.61 ± 0.36

* a significant difference is pointed out between the variants control and vibration + acceleration.
$\gamma = \gamma$-irradiation; v = vibration; a = acceleration.

To analyse the effect of dynamic factors on radiation damage the dry seeds were exposed before and after the action of dynamic factors to acute γ-irradiation by ^{137}Cs (3 krad, 517 rad min^{-1}). The interval between the action of dynamic factors and irradiation was not over 10 min. The effect was estimated by counting chromosome aberrations in root meristem cells of seedlings. The treated and control seeds were grown at 24 °C in Petri dishes on filter paper wetted in a 0.01% colchicine solution. The seedlings were fixed in alcohol ethanol acetic acid (3:1) and stained with acetocarmine. Aberrations in metaphase were counted on squashed preparations. The experiments were repeated three times. Since no differences between the replicates were detected the results have been combined.

The dynamic factors caused a statistically significant increase in the rate of chromosome aberrations (Table 3), mainly in the percentage of cells with chromosome aberrations (0.06% in the control and 1.20% in the experiment) and isolocus breaks (NUpd) (0.12% in the control and 0.62% in the experiment).

The comparison of these results with the data obtained earlier in flight experiments shows that dynamic factors as well as space flight factors increase the rate of chromosome aberrations. However, this increase is not always statistically significant in the material exposed on spaceships and much less than the effect of dynamic factors.

Modification of radiation damage in model and flight experiments is different. Thus, the space flight factors increased the radiation damage of seeds irradiated before the flight. In the experiment on the orbital station Salyut this effect was statistically significant. At the same time the exposure of irradiated seeds to dynamic factors does not change the radiation effect. The preliminary action of space flight factors did not change the effect of the following irradiation of seeds, whereas the action of dynamic factors in a similar experiment significantly decreases the radiation damage. The effect of radiation in the experiments on spaceships is possibly modified not only by dynamic factors but also by weightlessness. Further studies will show how true to the fact this suggestion is.

On the basis of our results it seems possible to make the following conclusions:

1. Dynamic factors (vibration and linear acceleration) increase the rate of chromosome aberrations in the meristem of *Crepis capillaris* seedlings. This increase is effected mainly through chromosome aberrations and isolocus breaks (NUpd).
2. A modifying effect of dynamic factors (vibration and linear acceleration) on radiation damage of *Crepis capillaris* seeds has been shown. The dynamic factors decreased the radiosensitivity of seeds to subsequent γ-irradiation and did not affect the effect of prior irradiation.
3. The results on the effect of space flight factors on dry *Crepis capillaris* seeds obtained in the experiments on spaceships cannot be accounted for by the action of dynamic factors alone.

References

[1] N. P. Dubinin, E. N. Vaulina, K. V. Kosikov et al., Life Sciences and Space Research XI, 105 (1973).
[2] N. L. Delone, V. F. Bykovsky, V. V. Antipov et al., Kosm. Issled. 2, 320 (1964).
[3] N. L. Delone, B. B. Egorov and V. V. Antipov, Kosm. Issled. 4, 156 (1966).

[4] M. A. ARSENIEVA, V. V. ANTIPOV, V. G. PERFUKHIN et al., Iskusstvennie sputniki Zemli, USSR Acad. of Sci. Publ. House **10**, 82 (1961).
[5] V. V. HVOSTOVA, S. A. GOSTIMSKI, V. S. MOŽHAEVA et al., Kosm. Issled. **1**, 186 (1963).
[6] YU. S. DIOMIN, Radiobiology **4**, 563 (1964).
[7] G. L. POKROVSKAYA, L. A. BELYAEVA and A. V. GOLOVKINA, Kosm. Issled. **3**, 652 (1965).
[8] M. A. ARSENIEVA, L. A. BELYAEVA and A. V. GOLOVKINA, Problems of Space Biology, IV, Nauka, Moscow 1965 (p. 373).
[9] E. V. MOSKVITIN, Annals of the 8th Working Meeting on Substance Circulation in a Closed System, Naukova Dumka, Kiev 1974 (p. 102).

Life Sciences and Space Research XIII — Akademie-Verlag, Berlin 1975

EFFECTS OF SPACE BALLOON FLIGHTS ON REPRODUCTIVE ACTIVITY IN *Paramecium aurelia*

H. Planel, J. P. Soleilhavoup and F. Croute

Laboratoire de Biologie Médicale, Université Paul Sabatier, Toulouse, France

Post autogamous *Paramecium aurelia* cultures were placed in hermetic containers, including a heating device with accuracy kept around $\pm 0.1\,°C$. Kinetics of cellular growth was determined by cell count, after recovery, on in-flight cultures and ground control cultures. Dosimetry was performed by thermoluminescent detectors ($CaSO_4$ activated with dysprosium). Flight durations of maximum altitude (ceiling) ranged between 48 min and 15 hours (repeated flights). Conclusions are as follows: short flights result in a secondary stimulating effect, shown by post-flight increase of the growth rate (total dose above 2 mrads); long flights or repeated flights are accompanied by a decrease in growth rate (total dose ranging from 2 to 6 mrads); in the stimulation experiments, cell counts performed immediately after flight permit identification of a temporary decrease of growth rate.

The biphasic character of the biological response after flights may be due to an ionization phenomena induced by cosmic rays. Indeed, the temporary drop of growth rate is not observed after recovery if the cells are subcultured in fresh medium and left on the earth's surface. We observe, on the contrary, an increase in growth rate.

These findings confirm the great sensitivity of *Paramecium aurelia* to very low doses of ionizing radiations and demonstrate the biological effect of cosmic radiation.

1. Introduction

In previous works we have shown that natural ionizing radiations may have an effect on living organisms. In particular, background radiation may exhibit a stimulating effect on the capacity of the unicellular organisms *Paramecium aurelia* and *Paramecium caudatum* to proliferate as shown by three pieces of evidence: (i) a decrease in the growth rate (Figs. 1, 2) or a lengthening in the generation time observed when cells are cultivated under natural or experimental shielding device [1—4]; (ii) cellular proliferation is normal when shielded cultures are simultaneously irradiated by very low doses of gamma radiations of thorium (Fig. 3) or ^{60}Co [2, 3]; (iii) on the contrary, when cultures are grown on the top of the Aiguille du Midi, in the Alps, at an altitude of 3800 m (Fig. 4), faster growth was observed as cultures were submitted to a higher level of cosmic radiation [5].

We have now studied the behaviour of cultures flown in stratospheric balloons and which have been irradiated by a high flux of energetic particles of primary and secondary cosmic rays.

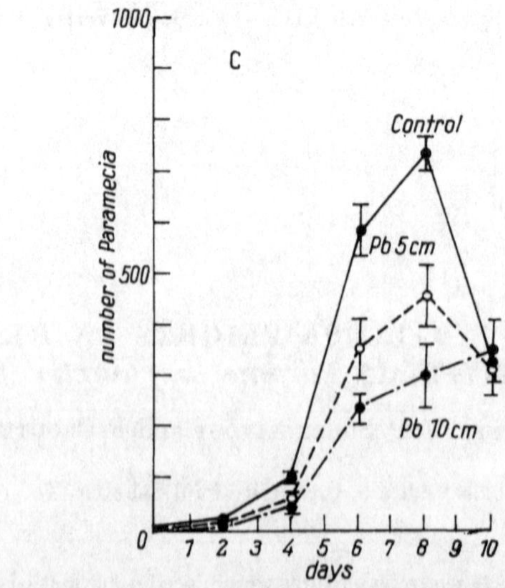

Fig. 1. Effect of shielding on the kinetics of cellular proliferation in *Paramecium*.

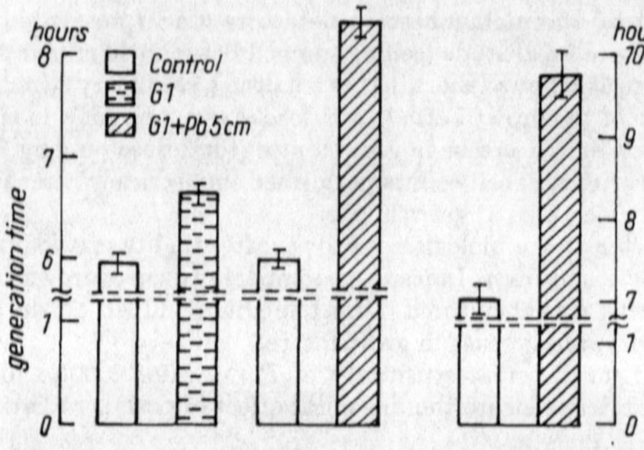

Fig. 2. Generation time of *Paramecium* cultivated in an underground laboratory (G1), and under a lead shielding (G1 + Pb 5 cm).

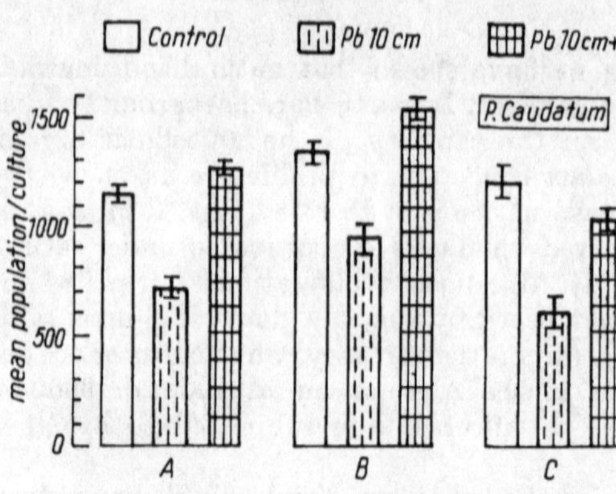

Fig. 3. Mean cell populations of *Paramecium* cultivated under a normal environment, in a lead shielding device (Pb 10 cm), and in an identical device with gamma simultaneous irradiation from a thorium source (Pb 10 cm + Th, dose rate ~1.9 mrad day^{-1}).

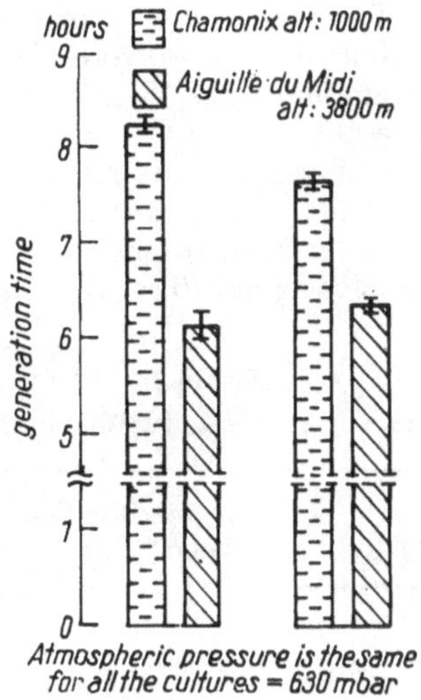

Fig. 4. Generation time of *Paramecium* cultivated at different altitudes.

2. Material and Technics

Paramecium aurelia (var. 2 Sonneborn Mating type VII) was used, 7 to 10 days after autogamy, i.e. in its most radiosensitive period [6]. Clonings were performed on strains radiosensitive or radioresistant to natural ionizing radiations [7] just before launching. Each culture consists of one cell and 4 ml of medium (Scotch grass infusion containing the bacterium *Aerobacter aerogenes*). For each flight, 235 cultures were set up with three radiosensitive strains and 156 cultures with

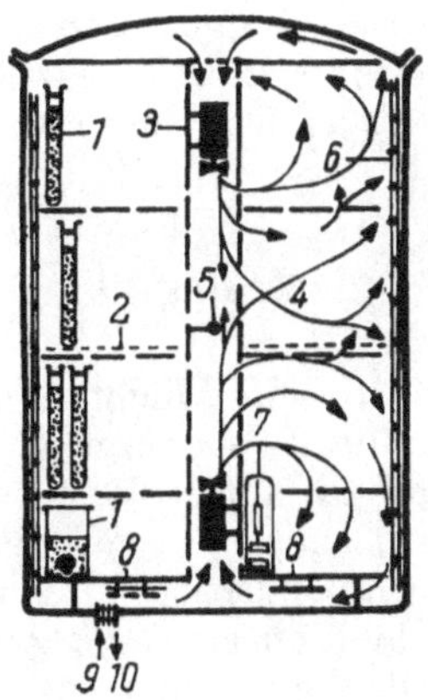

Fig. 5. Biological container used in the balloon flights.

1, Biological cultures; 2, Detectors; 3, Fans; 4, Air turbulence view; 5, Thermistor; 6, Heating element; 7, Potentiometric barometer; 8, Electronic circuits; 9, Energy supply; 10, Telemeasure.

two radioresistant strains. The same number of control cultures was kept on ground. Experimental and control cultures were placed in cylindrical 30 l. hermetic containers having aluminium walls (Fig. 5). A temperature of 25 ± 0.1 °C at every level was obtained by the use of a thermostatic heating device, combined with blown air; the accuracy of temperature regulation has been maintained in stimulation tests for outside temperatures ranging from +30 to −50 °C.

The nacelle consists of five containers enclosed within a thermo-insulation device made up of layers of polystyrene 10 cm thick and two layers of additional

Table 1
Duration of Flights and Radiation Doses

No. of Flights	Altitude (mbar)	Ceiling Duration (hours)	Accumulated Dose (mrad)
7	$\simeq 5$	$\leqq 5$	$\leqq 1.5$
7	$\simeq 5$	$\simeq 5-7$	$\simeq 1.5-2.1$
2	$\simeq 5$	$\simeq 7-9$	$\simeq 2.1-2.7$
4	$\simeq 5$	$\simeq 9-15$	$\simeq 2.5-4.5$

insulation. Energy is provided by batteries placed in the middle of the nacelle. Twenty flights were performed in 1972 and 1973 with launching at Gap or Aire/Adour (France). Ceiling durations range from 1 h to 9 h. In four cases, cultures were submitted to repeated flights and flew for 9 to 15 hours at ceiling (Table 1).

After recovery, flight and control cultures were placed in identical containers, at 25 ± 0.1 °C. Growth rate was determined by cell counts carried out every day from the second to the fifth day after cloning. About 20 cultures were tested in every flight. Cell counts were performed by a cell counter (Coultronics).

Dosimetry of balloons flights was performed by gamma spectrometry (Centre d'Etude Spatiale des Rayonnements de Toulouse) and thermoluminescent detectors (Service de Dosimétrie Physique, Commissariat à l'Energie Atomique, Fontenay aux Roses). Irradiation simulation experiments have been carried out with ^{60}Co, after cloning for 6 hours. The dose rate was about 9.3 mrad day^{-1}.

3. Results

1. If cultures are submitted to short flights (less than 6–7 hours), there is a slight but statistically significant acceleration of cell growth rate (Fig. 6). Nevertheless, this is preceded by a transitory inhibiting effect, shown by a decrease in growth rate: Fig. 7 shows the variations of cell growth relative to controls and it is seen that flown populations have at first a lower number of cells than controls. Later, this inhibition effect disappears and the growth rate of flown cultures rises and exceeds the controls.

2. If cultures are submitted to space flights with ceilings of long duration, more than about 7–8 hours, the stimulating effect disappears. Cellular kinetics is either normal or lower for 4–5 days after recovery for flights of 9 to 15 hours ceiling (Fig. 8). Variations in cell growth rate after balloon flights, estimated

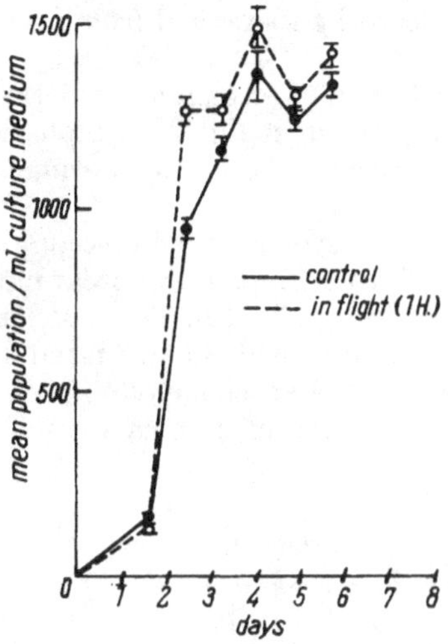

Fig. 6. Activating effect of a short duration ceiling flight in cellular growth rate. The vertical bars indicate the standard errors of the cell counts.

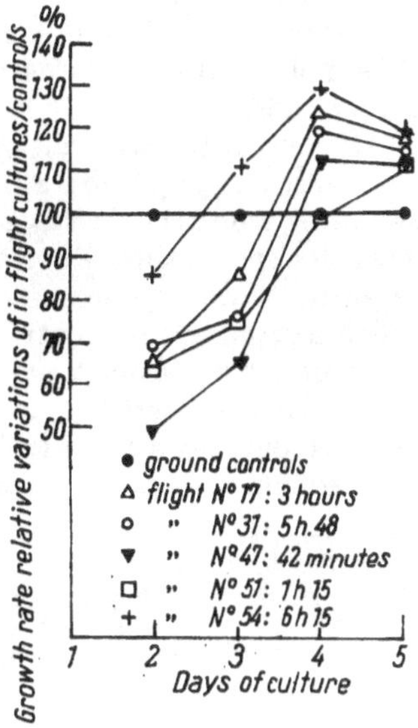

Fig. 7. Relative growth rate variations after a balloon flight. Biphasic biological response is shown.

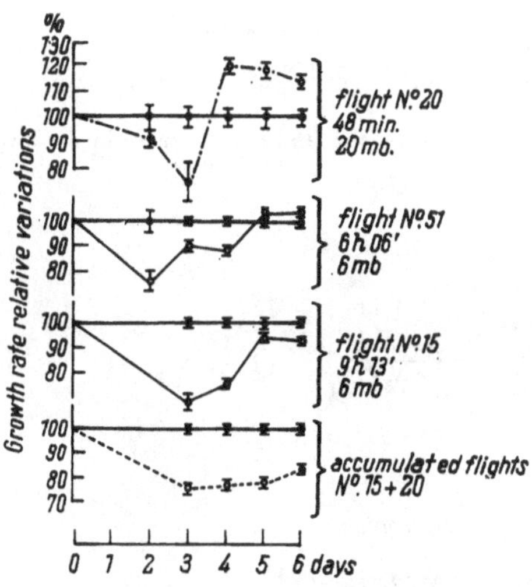

Fig. 8. Relation between variations of relative growth rate and the duration of flight.

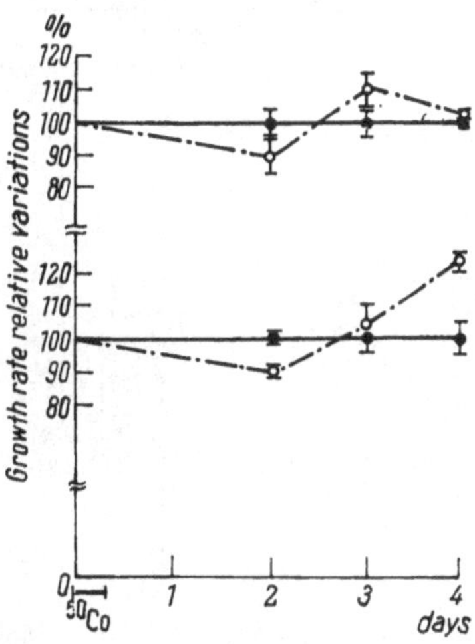

Fig. 9. Simulation, on the ground, of cosmic irradiation by ^{60}Co gamma rays (6 h at a rate of 1.23 mrad day^{-1}).

from cell counts, are statistically significant. Smaller or no stimulating effect was observed with radioresistant strains.

3. Accumulated dose of radiation during balloon flights ranged from less than 1.5 to 4.5 mrads.

4. Variations in relative cell growth after a low dose of gamma radiations of ^{60}Co (1.23 mrad day^{-1}, 6 hours irradiation) are plotted in Fig. 9. A biphasic response can be seen: the decrease in the growth rate is followed by a stimulating effect which appears three days after irradiation.

5. In order to determine the mechanism of the response of *Paramecium* after space flights or gamma irradiation, two lots of cultures comprising a great number of cells, were set up from a post-autogamous culture (Fig. 10). The first lot was exposed for 6 hours to a ^{60}Co source delivering a dose rate of about 1.23 mrad day^{-1}. The second group, used as control, was not irradiated. A specimen of uninoculated culture medium was also irradiated by the same dose of gamma rays. Later,

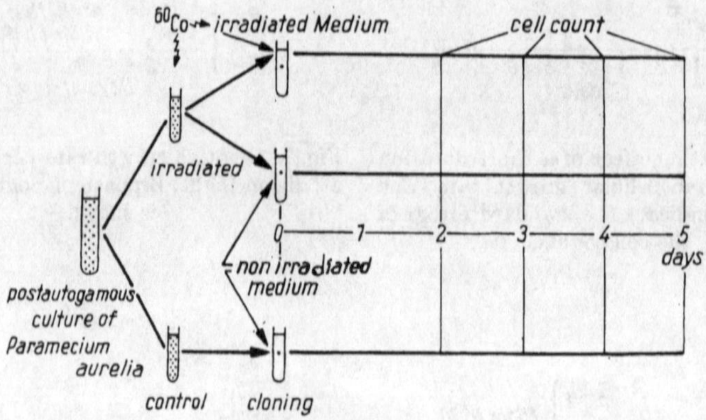

Fig. 10. General scheme of irradiation techniques of *Paramecia* and culture medium by low doses of ^{60}Co gamma rays.

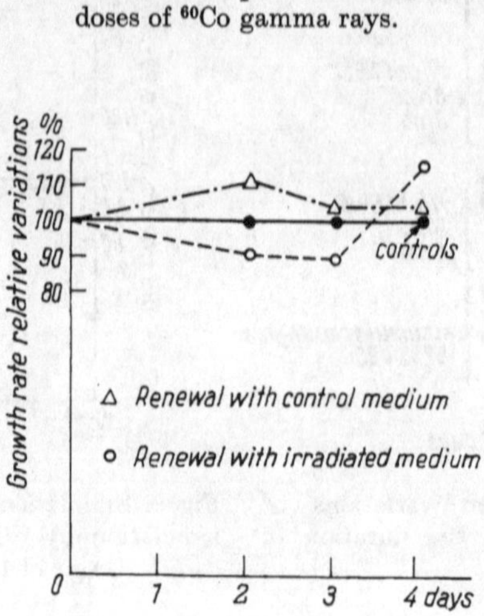

Fig. 11. Effect of irradiation of medium on cell growth rate.

cloning was performed from the irradiated culture, and in a first group of tests cells were individually incubated in the irradiated medium; in a second group, un-irradiated medium was used. Clonings were also performed from the control culture, using un-irradiated medium. Results are shown in Fig. 11: in the first group the cellular growth rate slows down at first and rises later, giving rise to a secondary stimulating effect; (ii) in the second group a higher relative cellular growth rate is observed, even in the earliest days.

4. Discussion

After balloon flights, a constant and significant increase in cellular growth rate has been observed in *Paramecium* cultures, when the ceiling duration did not exceed roughly 6 hours. It seems likely that this effect could be related to the increase in cosmic irradiation at increased altitude. Indeed, all other environmental factors were the same for in-flight and control cultures. Temperatures in particular were the same, both during and after the flight, as shown by telemetric and recording devices. The disappearance of all stimulating effects on cell growth rate for long duration flights suggests that this special effect of cosmic irradiation could be observed only for very low doses.

From dosimetric data a stimulating effect appears to occur only for a radiation dose lower than about 2 mrads. In other words, the stimulating effect of natural ionizing radiations appears to be limited to very narrow variations in doses. On the other hand, the flight experiments confirm previous studies, in particular the results obtained in our laboratory and at high altitude [5].

Nevertheless a stimulating effect on growth has not been observed in every study on space biology. Thus, proliferative activity of *Pelomyxa* [8] and of yeast cells [9] was not modified after a flight of short duration in a satellite. However, an activating effect has been described in *Escherichia coli* where the growth rate was increased after the Biosatellite 2 flight [10]. Furthermore, an activating effect has been described in vegetal material [11, 12]. In the experiment of Gertsusky [13] a stimulation was observed in seed germination and subsequent growth of the plants. In several species flown in Cosmos 110, the growth acceleration was explained by an increase in mitotic activity; in other words, a shorter generation time was observed and this can be compared with the biological response which appeared in our investigations.

The mechanism of this stimulating effect is still not obvious. It seems to be complex and probably variable in its time course, as is shown by the biphasic character of the biological response after space flight or irradiation. It seems likely that cosmic irradiation may first act through the culture medium. Indeed, experiments performed on ground have shown that when irradiated cells are immediately sub-cultured in a new medium, no inhibitory effect occurs, and only stimulation is observed. This suggests that, after an indirect effect of cosmic irradiation, a direct effect is produced which results in a stimulation of growth.

To sum up, these investigations show that cosmic rays may have a biological effect additional to the lethal effect of heavy ions as shown in Biostack experiments with *Artemia* eggs [14]. Ground and space experiments indicate that the nature of this biological effect can be affected by the accumulated dose.

References

[1] H. PLANEL, J. P. SOLEILHAVOUP and R. TIXADOR, C. R. Acad. Sci. Paris **260**, 3770 (1965).

[2] H. PLANEL, J. P. SOLEILHAVOUP, D. BLANC, J. FONTAN and R. TIXADOR, C. R. Acad. Sci. Paris **262**, 2767 (1966).

[3] H. PLANEL, A. BRU, J. P. SOLEILHAVOUP and R. TIXADOR, C. R. Acad. Sci., Paris **264**, 2945 (1967).

[4] H. PLANEL, R. TIXADOR, J. P. SOLEILHAVOUP and G. RICHOILLEY, C. R. Assoc. Anat. **149**, 951 (1970).

[5] J. P. SOLEILHAVOUP, F. CROUTE, R. TIXADOR, Y. BLANQUET and H. PLANEL, C. R. Soc. Biol. 1975 (in press).

[6] F. CROUTE, J. P. SOLEILHAVOUP and H. PLANEL, C. R. Soc. Biol. **166**, 191 (1972).

[7] F. CROUTE, H. PLANEL, J. P. SOLEILHAVOUP and Y. BLANQUET, C. R. Soc. Biol. **167**, 1984 (1973).

[8] J. H. ABEL, D. W. HAACK and R. W. PRICE, NASA Rep. SP 204, 1971 (p. 291).

[9] Y. G. GRIGORIEV and Y. Y. KOVALEV, NASA Rep. TTF 724, 1973.

[10] R. H. MATTONI, W. T. EBERSOLD, F. A. EISERLING, E. C. KELLER and W. R. ROMIG, NASA Rep. SP 204, 1971 (p. 309).

[11] L. K. GORDON, Kosm. Issled. **1**, 182 (1963).

[12] N. I. NUZHDIN, R. L. DOZORTSEVA and N. A. PASTUSHENKO-STRELETS, Izv. Akad. Sci. SSSR (Biol.) **4**, 576 (1965).

[13] D. F. GERTSUSKY, Kosm. Biol. i Med. **2**, 17 (1968).

[14] H. PLANEL, J P. SOLEILHAVOUP, R. KAISER and Y. BLANQUET, Life Sciences and Space Research XII, 85 (1974).

Life Sciences and Space Research XIII — Akademie-Verlag, Berlin 1975

THE EFFECT OF IONIZING RADIATIONS WITH DIFFERENT LET ON SURVIVAL AND MUTATION IN *Chlorella*

N. P. Dubinin, V. A. Shevchenko, A. V. Rubanovich, L. K. Vekshina and I. S. Sakovich

Academy of Sciences of the USSR, Moscow, USSR

The exposure of cells of *Chlorella vulgaris* LARG-I to radiations with a low LET (consisting of γ-rays) yielded sigmoid dose-effect curves, whereas the action of radiations with a high LET (multiply charged ions of carbon C_{12}^{+6}, boron B_{11}^{+2}, neon Ne_{22}^{+4}, argon Ar_{40}^{+8}) gave exponential curves. The rate of visible mutations depends directly on dose in the case of radiations with a low LET; it is a maximum after the action of radiations with a high LET. The highest RBE is shown by multiply charged ions of carbon (dE/dX 2850 MeV g^{-1} cm^2). With dE/dX higher and lower than that of carbon ions the biological efficiency of radiations is less.

1. Introduction

Studies of the effects of heavy charged particles are needed to solve problems of radiation safety during long-term space flights [1]. Difficulties in studying the action of heavy ions result from their low penetration (100—200 μm). *Chlorella* is a convenient object for investigation of heavy particle action since the cell diameter of this green alga is not over 10 μm. *Chlorella* is also a promising object for closed ecological systems of space vehicles. These facts stimulate interest in studying their radiation biology.

Work carried out with bacterial cells [1—6] has shown that heavy ions have lower biological efficiency compared with gamma radiation. This is also true for enzymes and viruses, but heavy ions had greater biological effects than γ-quanta on yeast cells and on mammalian cells in culture [8—10]. Thus, there is a difference between relative biological efficiency (RBE) of heavy ions depending on whether prokaryotes or eukaryotes are irradiated. We therefore have extended the study to *Chlorella,* which is a eukaryotic protozoon.

2. Methods and Material

Synchronous *Chlorella* cultures, strain LARG-I, were studied (cells were irradiated at the G_1-stage). The irradiation technique and data on the influence of UV-rays on *Chlorella* cells are reported in [6]. The cells were irradiated by γ-rays on the EGO-4 apparatus of the Institute of Biological Physics of the USSR Ministry of Public Health (dose rate 78.6 rad s^{-1}).

The cells were irradiated by multiply charged ions on the UMZI-300 cyclotron of the Joint Institute for Nuclear Research (Dubna). The work was carried out on apparatus specially constructed by the Institute of Medical-Biological Problems of the USSR Ministry of Public Health ("Argon" with electronic physical equipment). A monolayer of synchronized *Chlorella* cells was placed on the surface of nutrient-deficient agar in small Petri dishes (diameter 30 mm). The dishes were irradiated for different times at 100 rad s^{-1}. The error in the estimation of dose was not over 10%. Accelerated ions of boron (B_{11}^{+2}), carbon (C_{12}^{+6}), neon (Ne_{22}^{+4}) and argon (Ar_{40}^{+8}) were used in the experiments. The doses ranged from 0.5 to 30 krad.

The cells irradiated by heavy ions were then replaced in sterile distilled water and grown on agar nutrient medium to estimate cell survival by counting the microcolonies and the frequency of morphological and pigmental mutations [6].

3. Results

The irradiation of *Chlorella* cells by γ-rays yielded sigmoid dose-effect curves as was found before [15], whereas with multiply charged ions the curves of survival were exponential ones (Fig. 1). In irradiation by γ-quanta there is a linear increase in the rate of induced mutations with increasing dose (Fig. 2).

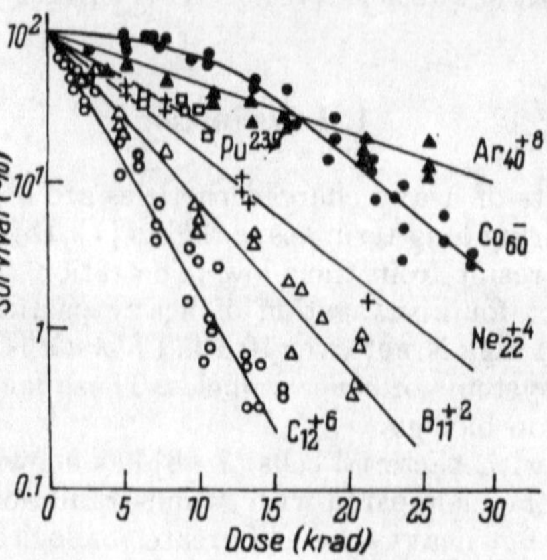

Fig. 1. Correlation between survival of LARG-1 *Chlorella* cells and the dose of ionizing radiations. The curve ^{60}Co is the sigmoid curve for γ-rays; the others are exponential curves for heavy ions.

Not a single experiment has demonstrated a decrease in the yield of mutations with increasing doses of γ-rays as is seen with radiation with a high LET (Fig. 2).

Coefficients of RBE of different kinds of ionizing radiations were estimated, relative to γ-rays. As the data in Fig. 1 show, the value of RBE depends on the dose (due to the sigmoid curve for γ-rays). Therefore the RBE was estimated for several different celle survival levels (10, 50 and 90%). On the whole, RBE coefficients of heavy ions are more than 1 (Fig. 3), i.e. the same relationship as observed for other eukaryotic organisms [7—10]. The highest RBE is shown by carbon ions; ions of boron and neon have less RBE. The maximal RBE for ions of LET $\sim$ 2000 MeV g^{-1} cm^2 has been shown for haploid and diploid yeasts [7, 11] and for mammalian cells [9, 12]. With LET over 3000 MeV g^{-1} cm^2 the RBE decreases, which proves that penetration of a heavy charged particle through

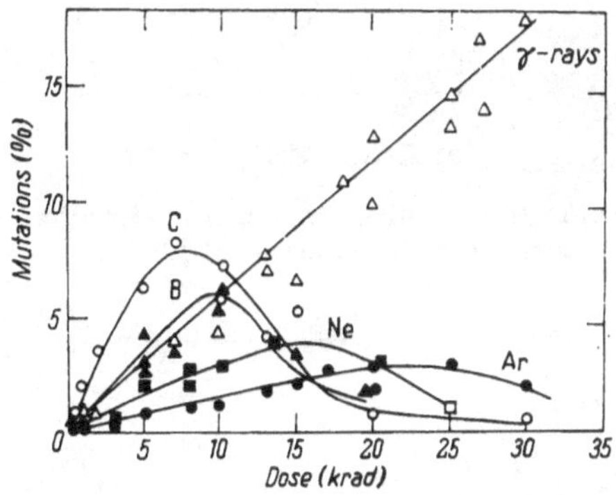

Fig. 2. The rate of visible mutations induced by γ-rays, ^{60}Co and multiply charged ions of carbon (C), boron (B), neon (Ne) and argon (Ar) in *Chlorella* cells.

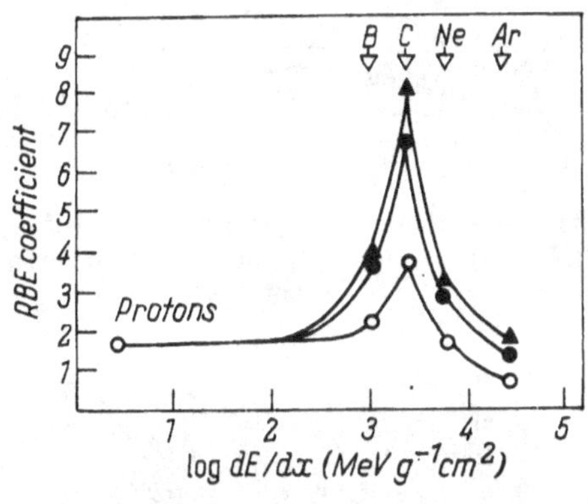

Fig. 3. The dependence of RBE coefficients on dE/dX of radiations for different levels of cell survival (▲—▲, 90% survival; ●—●, 50%; ○—○, 10%).

the sensitive volume leads to the loss of more energy than is required for induction of lethal effects. The highest RBE is observed at 90% level of cell survival, the lowest at 10% for all radiations.

The estimate of RBE coefficients based on induction of mutations was lower than that based on lethal effects (compare Figs. 3 and 4).

The effect of the maximal number of mutations which is observed after radiations with a high LET (Fig. 2) is of great interest. We compared the data on induction of visible mutations in *Chlorella* by ethyleneimine, UV-rays, γ-quanta, X-rays and multiply charged ions with their lethal effect (Fig. 5). The most effective mutagenic factor is ethyleneimine (the maximal number of visible mutations with a minimal lethal damage of cells), next come X-rays, γ-rays,

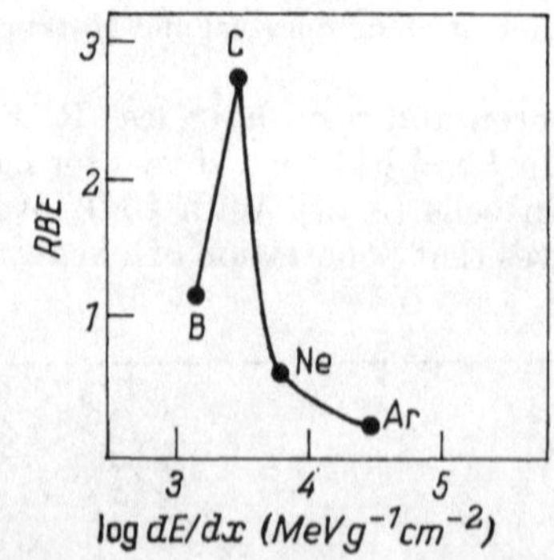

Fig. 4. Relative genetic efficiency of different kinds of radiation depending on dE/dX of radiation. The criterion is the induction of 1.5% of visible mutations.

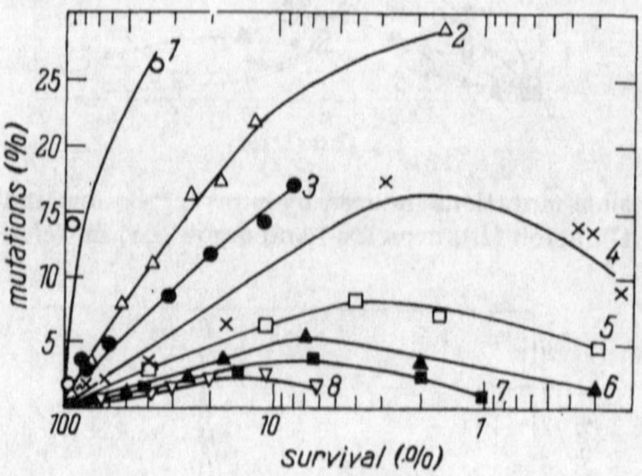

Fig. 5. Comparative mutagenic efficiency (with respect to the lethal effect) of ethyleneimine (1), X-rays (2), γ-rays (3), UV-rays (4), multiply charged ions of carbon (5), boron (6), neon (7) and argon (8).

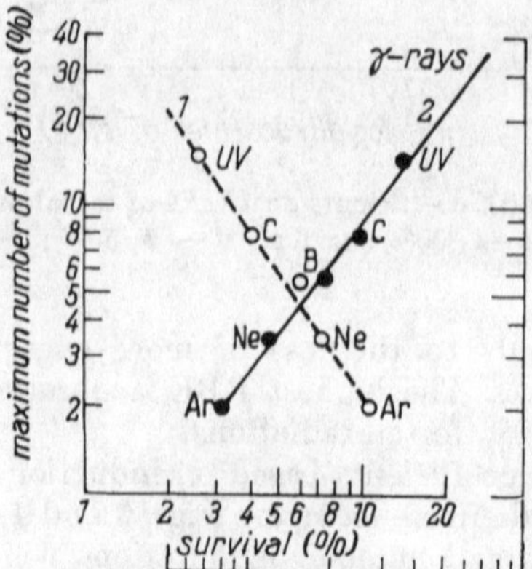

Fig. 6. Correlation of the maximal number of mutations with survival of the culture (1), and also with the number of mutations induced in *Chlorella* cells by different kinds of ionizing radiation (2).

UV light, and last the ions of carbon, boron, neon and argon. A particular correlation ought to be pointed out: the less the ratio between the number of mutations and survival at low doses (in the linear section of the curves in Fig. 5), the higher is survival upon which the maximum is observed. The maximal number of mutations is in good correlation with survival at which the maximum is observed (Fig. 6). The same figure shows the dependence of the maximum on the number of mutations at LD_{50} of the mutagenic factor in question. The data permit a preliminary estimation of the maximum dose of radiations for which we can estimate the number of mutations at LD_{50}. In particular Fig. 5 suggests that for X- and γ-rays the plateau or the maximal number of mutations should be observed at 1—2% cell survival and amount to 30—40% of mutations. (The maximal number of mutations is constant for each kind of radiation.)

4. The Nature of the Effect of the Maximum of Mutations

Interpretations of this phenomenon are given in a number of works (see the review [13]). Most of the hypotheses are object-specific and suggest that many mutations have higher radiosensitivity than normal cells. Such a situation can arise for various reasons: damage to cytoplasm, to repair systems etc. These can explain the maximum mutation effect. One would expect the phenomenon to have a simple and universal explanation. First of all, it is necessary to find the conditions for the existence of the effect. Such conditions can be simulated using the stochastic concept [14]. Leaving aside the mathematical aspect let us formulate the problem in simple terms: the necessary condition for the existence of the effect is a reduced mutability of irradiated cells. Let us give a more exact wording: if in the course of a radiobiological experiment we have a maximum of mutations with respect to surviving cells, the irradiated cells have mainly received irreversible damage which accounts for their reduced mutability. The above-mentioned damage can be called pre-lethal: any damage of the cell in the pre-lethal condition results in cell death. This can be expressed in the following inequality (much simplified):

$$q_{pl} < q_m + \alpha_m$$

in which q_{pl}, q_m are the intensities of cell death in the pre-lethal conditions and mutant cells respectively, and α_m is the intensity of repair of mutational damages. The order of magnitude of the dose that induces the maximum of mutations D_m is estimated by the formula:

$$D_m \sim \frac{1}{q_m + \alpha_m - q_{pl}}.$$

It follows from this that the maximum mutation effect should be expected for objects yielding exponential survival curves. For cells having sigmoid survival curves, e.g. for cells with a high repair efficiency, the maximum is achieved at very large doses. This conclusion is confirmed by exposure of *Chlorella* cultures to ionizing radiations with different LET (Fig. 5).

It would be interesting to detect the nature of pre-lethal damage. Irradiation with a low LET is known to induce mainly single DNA breaks, which in the course of metabolic processes are repaired to give the original structure, or else to give mutations. Irradiations with a high LET induce mostly double breaks which are most often not repaired at all. Therefore double breaks (or their combinations) may well be responsible for the pre-lethal damage. Cells so damaged cannot mutate, since any further damage causes their death. Mutations induced by small doses of multiply charged ions and associated with single DNA breaks (e.g. from secondary δ-electrons) are soon inactivated and after that this class of cells does not increase. These suggestions are supported by the fact that the maximum of mutations is observed with radiations with a low mutagenic efficiency but producing severe damage (Fig. 5).

5. Conclusions

Experiments were carried out on the effect of ionizing radiations with different LET (γ-quanta, multiply charged ions of carbon C_{12}^{+6}, boron B_{11}^{+2}, neon Ne_{22}^{+4} and argon Ar_{40}^{+8}) on cells of the green alga *Chlorella vulgaris* (strain LARG-1). It has been shown that survival curves are sigmoid in the case of radiations with a low LET, but with radiation with a high LET they are exponential. There is a direct correlation between the rate of pigmental and morphological mutations and the dose in the case of radiations with a low LET but there is a maximum for irradiation with a high LET. The analysis of the phenomenon of the maximum of mutations suggests that it occurs when radiation damages responsible for visible mutations and lethals are different, and when most damages causing lethals are irreversible.

The experiments have shown different efficiencies of different kinds of radiations inducing lethal damages and visible mutations. The highest RBE is shown by multiply charged ions of carbon (dE/dX 2850 MeV g^{-1} cm^2). When dE/dX is less or more than that of carbon ions the biological efficiency of radiation is lower.

References

[1] Yu. G. Grigoriev and E. E. Kovaliov, Physical and Radiobiological Investigations on board Spaceships, Atomizdat, Moscow 1971.
[2] R. Deering, F. Hutchinson and P. Schambra, Aerospace Med. **32**, 915 (1961).
[3] T. Brustad, Advan. Biol. Med. Phys. **8**, 161 (1962).
[4] R. Deering, Radiation Res. **19**, 169 (1963).
[5] C. A. Tobias and P. W. Todd (UCRL-11033), in: Semi-annual Rep. on Biology and Medicine Donner Lab., 25, 1964.
[6] E. A. Krasavin, thesis, University of Moscow 1971.
[7] J. A. Sayeg, A. C. Birge, C. A. Beam and C. A. Tobias, Radiation Res. **10**, 449 (1959).
[8] J. T. Lyman, Univ. California Lawrence Radiation Lab., Berkeley, Calif., AEC Contract No. W-7405-chg-48 UCKL-16030, 1965.
[9] P. Todd, Radiation Res. Suppl. **7**, 196 (1967).
[10] M. R. Raju et al., Radiation Res. **47**, 635 (1971).
[11] R. Mortimer, T. Brustad and D. V. Cormack, Radiation Res. **26**, 465 (1965).
[12] G. W. Barendsen, Nature, **193**, 1153 (1962).
[13] V. N. Soyfer, Molecular Mechanisms of Mutagenesis, Nauka, Moscow 1969.
[14] O. Hug and A. Kellerer, Stochastic Radiobiology, Atomizdat, Moscow 1964.

Life Sciences and Space Research XIII — Akademie-Verlag, Berlin 1975

PECULIARITIES OF BIOLOGICAL ACTION OF HADRONS OF SPACE RADIATION

I. G. Akoev and S. S. Yurov

Institute of Biological Physics, USSR Academy of Sciences, Puschino, USSR

Biological investigations in space enable one to make a significant contribution on high-energy hadrons to biological effects under the influence of factors of space flights. Physical and molecular principles of the action of high-energy hadrons are analysed. Genetic and somatic hadron effects produced by the secondary radiation from 70 GeV protons have been studied experimentally. The high biological effectiveness of hadrons, great variability in biological effects, and specifically of their action, are associated with strong interactions of high-energy hadrons. These are the probability of nuclear interaction with any atom nucleus, generation of a great number of secondary particles (among them, probably, highly effective multicharged and heavy nuclei, antiprotons, π^--mesons), and the spatial distribution of secondary particles as a narrow cone with extremely high density of particles in its first part. The secondary radiation generated by high- and superhigh-energy hadrons upon their interaction with the spaceship is likely to be the greatest hazard of radiation to the crew during space flights.

1. Introduction

A review of experimental data is presented on the somatic and genetic effects of high energy hadrons with discussion on physical and molecular principles that are responsible for the biological effects of hadrons, and the data from high energy accelerators and space studies.

2. Analysis of Biological Experimental Data obtained in Space

The main results from available information on biological studies in space performed by Soviet and American scientists [1—3] are as follows:

(i) Biological effects vary widely, and there is no distinct dependence of somatic and genetic effects (when they were observed) on the radiation dose, on the duration of weightlessness or on other factors accompanying space flight (FSF). It is also difficult to reproduce the experimental results under conditions of space flight or those similar to it. In addition, one may observe different biological effects in replicate samples in the same experiment.

(ii) There is no correlation between degree of radioresistance and the effect of FSF. The biological materials in these experiments differ tenfold in radiosensitivity. However, FSF may reduce the ability to repair genetic and somatic damage.

(iii) It is possible that space factors produce high biological efficiency at low radiation doses. There is great discrepancy between biological effects of radiation in space and on the earth. However, this is not explained by a combination of radiation and weightlessness. This is shown by the experiment on sputnik "Bios-II", in which an internal radiation source was used, because the effect of weightlessness was found to be slight or absent.

(iv) Factors accompanying space flight have certain biological actions, seen in genetic and morphological effects absent on the earth (such as formation of spherical segments in chromosomes, unusual disturbances in cell division and defects in embryonic development, scattered microscopic lesions in tissue, multiple damage in hereditary structures) or in a substantially changed ratio of types of chromosome damage.

The above mentioned peculiarities of biological studies in space made us [2, 3] assume that there is a factor responsible for them which: (a) is not taken into consideration in biological experiments in space or is interpreted incorrectly: (b) has considerable biological efficiency: (c) possesses a certain specificity in its biological action: (d) is transiently effective at each point within the volume of the biological object. The super-hard component of cosmic radiation (above several tens GeV), represented by high energy hadrons, may be such a factor. Hadrons of high and superhigh energies form only a small part of the whole flux of cosmic radiation. Nevertheless high biological effectiveness may be expected owing to the features of strong interaction peculiar to high energy hadrons [1—3]. It may be expected that the interaction between the body of the ship and primary protons from the galaxy, or other hadrons of superhigh energy, would give rise to a narrow cone of secondary radiation containing multicharged nuclei, hadrons of different kinds and charges together with more usual radiation, all of which would influence the crew and biological objects. Because of this an extremely irregular absorption of energy is expected in minute volumes which are some orders of magnitude less than those of the ionizing chambers and activating detectors customarily used. That is why only an averaged dose is usually measured rather than that really absorbed by single microscopic objects [9].

These observations may well be explained on the assumption that concurrent with the usual cosmic radiation there are hadrons of high and superhigh energy acting, and contributing their own statistical regularities. These comprise strong interaction and specific biophysical mechanisms responsible for high biological efficiency, substantial variability of biological effects and a certain degree of specificity.

3. Results of Biological Studies with Accelerators

Studies of the biological effectiveness of high energy hadrons were started on the Serpuhov 70 GeV proton accelerator. Dosimetry and technique of the biological experiments have been described earlier [3]. The region of the major targets on the accelerator was chosen for the experiments, because it was here that the mixed nuclear-active flux of secondary radiation was generated. This permitted estimation of the danger for radiation risks in work on the accelerator and also for purposes of cosmic biology, because in a space ship the mixed nuclear-active cone of secondary radiation (formed by the primary particle of superhigh energy

Table 1

Summary Table of RBE Data of Secondary Radiation from 70 GeV Protons compared with the action of γ-Radiation [1–3].

Criteria	Mean RBE values from approximate curves	RBE_M
Physiological data for *Vicia faba*		
Survival, length of the primary root and dry weight of seedlings	1.63–1.85*	2.03–4.61**
Repair rate for growth of the primary root	1.85	—
Chromosome aberrations of *Vicia faba*		
Number of aberrant cells with fragments or bridges or both	1.52–1.91*	4.70**
Number of fragments or bridges or both per cent of aberrant cells	1.84–2.20*	3.20–4.50**
Rate of decrease of aberrant cell number	1.71–1.74*	—
Rate of decrease of chromosome fragment number	1.54	—
Survival of *Escherichia coli* and bacteriophage T4B r^+		
E. coli B, *E. coli* K-12 p-678 in medium M-9 without glucose	0.8–5.1*	up to 10.0**
T4B r^+ in nutrient broth	0.5–4.2*	up to 34.0**
T4B r^+ in buffer	0.4–4.6*	up to 7.5**
Mutagenesis of bacteriophage T4B		
Frequency of r-mutants in broth	5.5	up to 12.5
Frequency of r-mutants in buffer	4.5	up to 13.0

* RBE values obtained by the criteria mentioned in [1–3].

** For the curves of an inverted type derived for small doses (calculated according to recommendations of ICRP for the curves from individual groups of experimental points).

when interacting with the material of the ship body) is the most important factor [5].

Hadrons of average energy between 10 and 40 GeV were the major constituents of the secondary radiation produced by the accelerator [1, 4]. Hence, the increased biological efficiency of secondary radiation of the accelerator is ascribed to effects of high energy hadrons. The averaged results given in Table 1 show that biological efficiency of hadron radiation can be high. The highest relative biological effectiveness (RBE) values were found with survival and mutogenesis of bacteriophage. The RBE of secondary radiation of the accelerator proved to be higher than was expected from traditional ideas used in setting radiation safety, i.e. that there is an unambiguous relationship between biological efficiency of radiation and the averaged values of linear energy transfer (LET) (Fig. 1). A low biological efficiency [6] was expected for corpuscular radiations of high energy (more than 1 GeV), since they have characteristic low averaged LET values of about 0.2 to 0.3 keV μm^{-1} [7, 8].

As already known [9], the RBE for various kinds of radiation decreases with a decrease in average LET values to below 100–400 keV μm^{-1}, irrespective of the criterion of biological action (Fig. 1) which is observed (in particular, as the proton energy rises from several keV to some hundreds of MeV). Here the RBE value becomes less than 1.0 [9]. This was assumed to be true for further increases in energy of protons and their associated secondary radiation. Indeed, as was shown at the Brookhaven laboratory [6] the RBE of 8 GeV μ-mesons equals 0.7, as

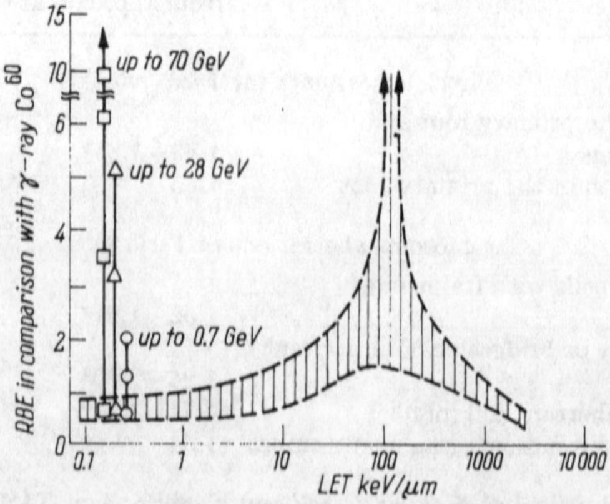

Fig. 1. Dependence of relative biological efficiency on average linear transfer of energy.

predicted. However, pi-mesons of the same energy and 28 GeV protons with the same LET values (0.25 keV μm^{-1}) possess a nearly 6-fold higher biological efficiency. Such an increase in RBE of high energy radiation has been observed by a number of authors [6, 8, 10] under conditions where the contribution of hadron effects was pronounced on top of the background of the usual ionizing radiations with their electromagnetic actions. Another feature of the experimental results obtained with the secondary hadron radiation on the Serpuchov accelerator is the variability of biological efficiency of radiation observed in a series of experiments, within one experiment and even at one experimental point (Figs. 2, 3). The greatest variability within experiments is seen with bacteria and bacteriophages (Table 1). The variations within one experiment (Fig. 2) were as if the irradiation doses of single experimental groups unexpectedly fluctuated by a factor of 2–5 fold increase. Though this was also observed with bacteriophage survival and chromosome aberrations of *Vicia faba*, these variations in radiation dose were not detectable by the activational detectors and ionizing chamber. Appropriate conditions were met for measuring radiation with the experimental groups, and magnetic field effects were probably eliminated [11]. Analogous variability of single groups of experimental points within the same experiment was observed in both the spectrum of r-mutations of bacteriophage T4B and the reversions of rII-mutations induced by accelerator radiation compared with the action of chemical mutagenes (the revertability varied from 0 to 70% in 11 experimental groups within one experiment). Variability was also observed for both single groups of experimental points within one experiment (Fig. 3B) and for one experimental point (Fig. 3A), in which several strains of bacterial and bacteriophages were studied simultaneously.

There are other effects which may be due to hadrons, as they are not described for any other kind of radiation. An example is the spectrum of bacteriophage

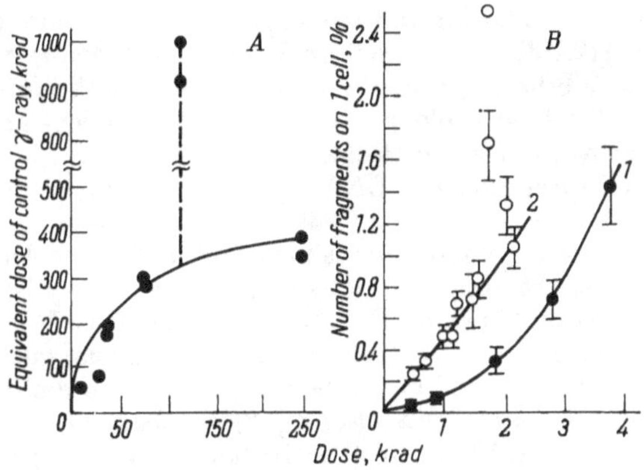

Fig. 2. Patterns of experimental spike points for the effect of the secondary radiation of 70 GeV protons on bacteriophage and *Vicia faba*: A, survival of bacteriophage T4B r$^+$ upon radiation in nutrient broth; B, frequency of chromosome fragmentation per aberrant cell of *Vicia faba* (data from B. S. Fomenko).

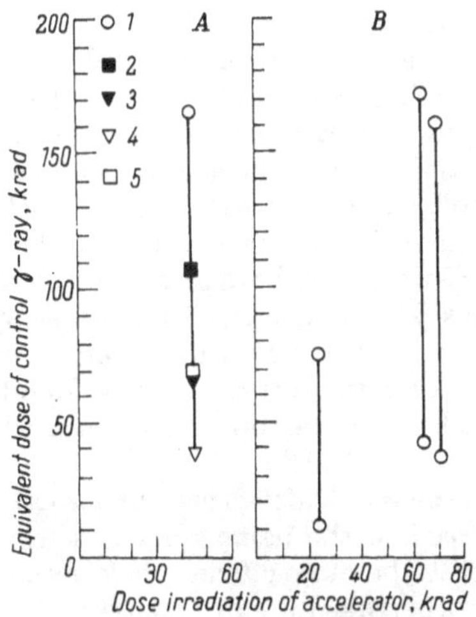

Fig. 3. Variability of biological effect. A: at one experimental point, in trial No. 3: 1, *Escherichia coli* K-12 P-678 leu$^-$, thr$^-$, B_1^-, Str^r; 2, *E. coli* B; 3, *E. coli* K-12 (λ) (all bacteria are in medium M-9 without glucose); 4, bacteriophage T4B r$^+$ in broth; 5, bacteriophage T4B r$^+$ in buffer. B: in the same experiment for different experimental points in trial No. 2: for bacteria *E. coli* K-12 P-678 leu$^-$, thr$^-$, B_1^-, Str^r.

mutations induced by accelerator radiation; this was most different in the number of mutations of a bacteriophage radiated in buffer. In two experiments their number was respectively 12 times and 15 times as high as in the control, whereas rII- and rIII-mutations were few in number. At the same time, the revertability of rII-mutants induced by a number of chemical mutagenes was different (revertability capacity was between 0 and 70%), whereas the revertability of rII-mutants induced by γ-radiation in these experiments was close to that obtained by other authors [12, 13]. It is a well known fact that non-revertability of rII-mutants induced by other agents may be evidence that the primary changes in the bacteriophage DNA molecule involve deletions, inversions and translocations [14]. Such damage to the DNA molecule may be associated with a nuclear interaction between a hadron and the DNA molecule, leading to unusual molecular damage.

Based on these data an assumption is made [2] that the molecular mechanism of mutagenesis under the action of high energy hadrons may be due to multiple DNA changes caused by "burning" by a narrow cone of secondary particles and, hence, production of deletions of various dimensions (from small to large ones). The secondary radiation turned out to be most efficient in multiple disturbances of chromosomal structures of *Vicia faba* (Fig. 2) per aberrant cell and, at the same time, the least efficient in the degree of activation of recovery processes using the same index. Hence, hadron radiation possesses not only a high biological efficiency but also a characteristic specificity seen in considerable variability in biological effects.

4. Biophysical Analysis of Hadron Effects

The experiments with an accelerator with 70 GeV proton secondary radiation basically consisting of high energy hadrons showed essentially the same peculiarities of biological action (high biological efficiency, substantial variability and the same specificity of biological effects) as those observed in studies in space. When analysing the biological basis for the effect of high and superhigh energy hadrons it was shown that the following features of strong nuclear interactions peculiar to high energy hadrons may be responsible for the biological effects observed: (i) probability of nuclear interaction with any atomic nucleus, as there are practically no resonance spikes of interaction; (ii) multiple generation of secondary nuclear-active particles, among which multiply charged and heavy nuclei, mesons, baryons, antiprotons and other antiparticles may be found; (iii) spatial distribution of these secondary particles as a narrow cone within which the particle density is extremely high.

High biological effectiveness of hadrons may be associated with the high density of energy in the initial part of the narrow cone; neutrons of 0.4 GeV from the CERN accelerator proved to be more efficient before the neutron absorber rather than after it [8]. The absorber was intended to lower the neutron energy with the aim of obtaining LET in the zone of the Bragg spike, due to production of nuclear spallation of a "star" type. The neutrons weakened by the absorber became less efficient because they could no longer generate narrow cones of secondary particles. The primary region of the narrow cone of secondary particles is biologically efficient because biological objects differing ten times in radiosensitivity do not exhibit any differences in disturbances of their FSF (in those cases in which these disturbances were observed at all). Recent studies on the sputnik Intercosmos 17 showed that a cosmic particle of extremely high energy ($\geqq 10^6$ GeV) is capable of generating many hundreds of secondary particles which retain the direction of the primary particle with an extremely small distribution angle of only a few degrees.

The considerable variability in the biological action of hadrons is due to an extremely non-uniform distribution of the adsorbed energy and this is associated with only a small probability that the primary region of the narrow cone of secondary particles will reach the biological object. Statistically random events (such as strong interactions, generation of secondary particles, production of multicharged and heavy nuclei of high efficiency, antiprotons and pi^--mesons) contribute to that variability. Microscopic disturbances in individual cell structures can be due both to multicharged and to heavy nuclei which yield an ionization track of about 50 Å diameter in the primary region of the secondary particle cone [15]. However, microscopic disturbances extending over numbers of cells, as well as morphological damage ("meltings") to the tissue are associated with the gradual expansion of the cone. With increasing hadron energy the probability of strong interaction per volume unit decreases. However, if the strong interaction occurs, its biological effects becomes greater, with a simultaneous increase in variability of the effects.

Until recently radiational biology and medicine dealt essentially with the effects of the radiations producing electromagnetic interaction and elastic nuclear interactions. Hadron biology has now made its first steps. It is becoming clear, however, that hadron biology is of importance not only in space studies and in radiation safety measures for new accelerators of superhigh energy, but also for observations made under Alpine conditions, or during civil aviation flights. All these considerations point to the importance of studying hadron effects [16—18].

References

[1] I. G. Akoev et al., Life Sciences and Space Research XI, 225 (1973).

[2] I. G. Akoev and S. S. Yurov, Preprint "On biophysics of strong interactions of high energy hadrons" T-12552, Puschino, USSR, 1973.

[3] I. G. Akoev and S. S. Yurov, Izv. Akad. Nauk SSSR, ser. biol. **1**, **11** (1975).

[4] I. G. Akoev et al., in: Proc. Int. Congr. on Protection against Accelerator and Space Radiation, CERN, Geneva **1**, 122 (1971).

[5] L. V. Kurnosova et al., Kosm. Issled. **9**, 105 (1971).

[6] H. H. Smith, Radiat. Res. suppl. **7**, 190 (1967).

[7] F. P. Cowan, Radiat. Res. suppl. **7**, 1 (1967).

[8] J. Baarli et al., in: Proc. Int. Congr. on Protection against Accelerator and Space Radiation, CERN, Geneva **1**, 133 (1971).

[9] N. G. Darenskaya, L. B. Kosnova, I. G. Akoev and G. F. Nevskaya, RBE of irradiations. Factor of time, Atomizdat, Moscow 1968.

[10] V. V. Shevchenko, Genetica **5**, 115 (1965).

[11] I. G. Akoev et al., Radiobiologiya **14**, 374 (1974).

[12] S. L. Phillips, S. Person and H. P. Newton, Int. J. Radiat. Biol. **21**, 159 (1972).

[13] H. V. Malling and F. J. de Serres, Radiat. Res. **53**, 77 (1973).

[14] M. Nomura and S. Benser, J. Molec. Biol. **3**, 684 (1961).

[15] C. A. Tobias, T. F. Budinger and J. T. Lyman, Preprint LBL-1011 AEC W-7405-eng-48 (1972).

[16] V. I. Danileiko, in: Problemi Kosmicheskoi Biologii **1**, Nauka, Moscow 1962 (p. 392).

[17] W. Haymaker et al., Aerospace Med. **41**, 989 (1970).

[18] I. G. Akoev et al., Kosm. Issled. **12**, 617 (1974).

INDEX OF AUTHORS

MIX
Papier | Fördert
gute Waldnutzung
FSC® C083411

Zeitfracht Medien GmbH
Ferdinand-Jühlke-Straße 7
99095 Erfurt, Deutschland
produktsicherheit@kolibri360.de